Kristopher Justin

Copyright © 2022 Kristopher Justin

All rights reserved.

DEDICATION

For Samantha, who always told me that I start way too many projects and never finish any of them. Gotcha this time. If it wasn't for your support, I never would have finished this one, for sure. You are the love of my life, and that is one thing that will outlast even the end of the world.

CONTENTS

	Acknowledgments	i
1	Introduction to the Wasteland	1
2	The Hard & Fast Limits to Growth	5
3	The Crisis of Climate Change	18
4	The Future of Pandemics	56
5	The Conflict of Nations	68
6	The Peak of Complexity	136
7	It All Comes Together in the End	148

ACKNOWLEDGMENTS

First and foremost, this project would not have been remotely possible without the support of the love of my life, Samantha. How she put up with all my messes, I will never know, but I will always be incredibly thankful that she did.

I would also like to thank many of my fellow Redditors on r/Collapse, some of whom were encouraging all through the writing process, others that helped expand my own understanding of collapse, and even those whose ignorance of the issues demonstrated the dire need for a book such as this. I am quite sure each of you knows who you are.

Finally, I would like to give a special recognition to Nate Polson from Canadian Preparedness. Known as "Canadian Prepper" on YouTube, he is the one who kept me updated with the relevant daily news during the writing of this book and saved me the time of having to sort through the mainstream media junk to find it. Not so much the climate stuff, perhaps, but all the rest. That's a hint, Nate. Overall, great job, and thanks.

1

INTRODUCTION TO THE WASTELAND

Wasteland by Wednesday. What does that mean exactly? Well, for one, it's where I think civilization is headed. A total societal collapse into an environment more often detailed in fiction than in any realistic projections. A wasteland of the decaying remains of a system that could no longer sustain itself through infinite growth in the face of dwindling resources and now cratered with the damage caused by its people during the descent. As for Wednesday, that's because I believe it stands a pretty good chance of happening much sooner than most others expect. And I also needed a catchy name for both my website, and this book, so there it is.

Another prepper thing, really? Not really. Well, kind of. While there are many elements of prepping to be found on my website of the same name, this book is not intended to be of the traditional prepping sort, or any prepping sort really. Most traditional prepping tends to focus on getting ready for local, or even regional, disasters, as opposed to a real 'End of the World' type of scenario. Prepping for the most likely occurrences is a good thing, and there is already plenty of information about it out there to be found, much of it written by those more knowledgeable than I. The direction that I intend to take here is to look at what sort of collapse can happen to society, and the process that I see as…maybe not most likely, but certainly possible and often ignored.

My purpose here is to inform you, and hopefully other readers, about the potential for a complete and catastrophic societal collapse on a global scale. Rapidly, and coming in the near term.

Like, the whole shebang, maybe even the last shebang. This book is not really about prepping and surviving, it is more of a different view of how things that are playing out in the world today can result in the sort of collapse that is usually relegated to the post-apocalyptic fiction shelves. What I am trying to do here is write a book that will showcase the big picture and introduce these truths to those who either are unaware or who simply do not understand it all. Before anyone can prepare to try and survive it, they have to know exactly what "it" is, right? And that is my goal.

Ambitious, right? And I am hardly an expert. In fact, it was my own searching for an expert that led me to write this book and create my website, as it would seem that no such expert exists. At least not in the form I was thinking. On one hand, there are a large number of very good authors who have written about the climate crisis and how it could lead to ecological collapse. And there are also those that write very well about the potential of other factors, like conflict, to drive us toward a social collapse. But, oddly enough, there do not seem to be very many that crossover between the many different factors and show how the gradual effects of climate change on humanity will put ever increasing pressure on our complex systems leading to a societal collapse well in advance of the ecological one.

Each person out there seems to have a dedication to the problem, or at least to the problem they have championed. Whether it is pollution of the oceans, deforestation, economic devastation, climate change, or even the potential for world war, there are many great people in the world fighting for change, most of them so much more knowledgeable than I am about it. But they all seem to miss the big picture that there are more threats than just one or two. Having a laser focus on fighting the destruction of the Amazon rainforest is good, but then we are divided because it falls to others to fight the overfishing of our oceans, or the expansion of the fossil fuel industry, or the increasing drives for conflicts between nations. Everyone seems to be doing what they can to bravely fight the battles, but no one is really fighting the overall war.

Another thing is that, while there is a wealth of scientific evidence and writings out there regarding the coming collapse, most of it is very complex and, let's face it, hard to read. Hard if you are just a normal person like me, that is. Without the intense interest to motivate me to learn everything I could about it, I never would have picked of some of the most important books out there to read, such as The Limits to Growth. And I certainly would not have taken the time to expand my education enough to fully comprehend all of it.

Finally, I have also noticed a distinct dichotomy amongst the various "types" of people out there who are keenly aware of some factors while almost entirely ignoring others. The scientific types tend to not look too closely at the non-science factors that contribute to collapse. You can see this both in their writings and in their interactions on various online forums like Twitter and Reddit (r/Collapse, I'm lookin' at you). While being intricately caught up in the science of climate change, deforestation, rising ocean acidity and all of those things, they pay almost no attention to the role of geopolitics, economics, or outright warfare and how those things will contribute greatly to collapse. Lip service is generally paid to the hardships those things cause, but rarely are they considered as primary drivers of collapse, or as the results of increasing climate pressures, and so things like the possibility of nuclear war are almost scoffed at. Intellectuals will discount such things as virtually impossible, since starting a nuclear war is, in their minds, completely irrational and not something any sane person would ever do.

Now, I hate to be the one to break it to them, but not all world leaders are sane and rational, and for some the very definition of rationality might be quite different. It is generally accepted among the intellectuals that humanity will just peacefully endure a long, slow decline in quality of life and availability of resources, and in the end, everyone will come together to help each other adapt to changing reality by working together with others in their community, and indeed the world. Obviously, these people have not tried to make their way through Walmart on Black Friday opening when there are only 20 PlayStation 5's left in stock. That is where you will see real human reaction to scarcity and disappointment in life.

The same goes for the other side of the spectrum as well. From the "crypto-bro's" to the financial analysts out there, concerns over the environment or climate are interesting, but those really don't have much to do with this quarter's revenue numbers, right? We can worry about that later, we know it is a very pressing issue, but right now "Oh my God, the economy!"

Then you have the prepper-types, of which I am one, at least the majority of the traditional ones. Ironically, those who are doing the most to prepare for plagues and invasions, civil wars and hurricanes, they end up either ignoring or outright disbelieving the much bigger threats of climate change which are the primary drivers of the events that they are preparing for.

No, what is needed at this point in the timeline for collapse is a quick layman's guide to all the factors out there that are driving us towards our doom. Something that can bring the sides together, that shows not just the threats that are bearing down on us, but how those threats are worsened and accelerated by each other. Something that tries to explain, in simple terms, the different perspectives of what we are all talking about, and that does it in a way that a regular human like me can understand it.

And so, here is my own attempt at that. I am quite sure that I am not as smart as the experts, but I am wise enough to recognize that fact and just intelligent enough to understand what they are saying. Putting those together, and adding in my own experience, is where my analysis and prediction comes from. Keep in mind that this book is not intended to be an intellectual exercise, nor is it meant to breakdown all of the science behind things, ad nauseam. It is a primer, to help readers get a glimpse of all of the crises of our current times, and to help them by acting as springboard to do their own in-depth research. That being said, my own thoughts on the material are not what really matters here. What matters is the information itself, which will help you understand what the world is facing from all angles, and help you put together your own informed ideas about where it all might lead. "Collapse enlightenment" is what I am going for here. So, let's get to it.

2

THE HARD & FAST LIMITS TO GROWTH

It is no big secret these days that the world as we know it is heading towards a collapse. If you look around you closely enough, you can see it everywhere, reflected in every facet of our daily lives. We actually have a roadmap for some of it, one made back in 1972 at the Massachusetts Institute of Technology, by a group of researchers way ahead of their time. Concerned about the direction the world was headed even back then, they got together to study the risks of civilizational collapse on a scientific level and see just what the chances of such a thing really were.

That MIT study and the book that followed, The Limits to Growth, originally published by the Club of Rome, was a very in-depth and comprehensive piece of work. Using their system dynamics computer model (called World3) they created a simulation to get an idea of where civilization was headed should humanity continue doing things as it had always done them. The result that emerged brought to light impending physical limits to possible growth that, when reached, meant modern industrial civilization was facing a total societal collapse around the mid-21st century. This would be primarily due to overexploitation of the Earth's natural resources in the pursuit of economic expansion.

That analysis generated heated debate even among the scientific community and was widely derided at the time by various pundits and talking heads who misrepresented its findings and methods greatly, much in the same way that the media continues to do today with every other warning since. Political and business interests in particular did not give it the slightest bit of credibility, and certainly didn't want anyone else to take it seriously.

That original study is pretty wild stuff, but even that dire prediction from so long ago is not the scariest part.

You see, just recently, that study was gone back over, given a detailed second look. Just to investigate how we were doing. Gaya Herrington, who is the Sustainability and Dynamic System Analysis Lead at KPMG in the United States, undertook that new project. Her new research study, done as a personal project on her own time, represents the first time a highly ranked analyst and executive working within a large multinational corporate entity has taken the old MIT "limits to growth" model seriously.

"Given the unappealing prospect of collapse, I was curious to see which scenarios were aligning most closely with empirical data today." Herrington says in explaining her project. "After all, the book that featured this world model was a bestseller in the 70s, and by now we'd have several decades of empirical data which would make a comparison meaningful. But to my surprise I could not find recent attempts for this. So, I decided to do it myself."

Somewhat like my own motivation to write this book.

So, she did it, as a part of her master's thesis at Harvard University. Just to get a feel for how well the MIT model stood the test of time, in her capacity as an advisor to the Club of Rome. And guess what? It seems like we have been following along pretty well, so much so that we are actually right on track to suffer such a collapse with perfect timing. Now that is scary.

This new look at the old study, titled "Update to limits to growth: Comparing the World3 model with empirical data," attempts to assess how MIT's World3 model measures up against all the new data available. It was published in the Yale Journal of Industrial Ecology in November 2020. In her research, Herrington comes up with some very startling findings. Her work concludes that the current "business-as-usual" trajectory of globalized civilization is heading toward the terminal decline of economic growth within the coming decade - yes, this decade - and could end up triggering total societal collapse by the year 2040 or so. The model has held true, and we have been cruising along, right on track, for this entire time.

This new analysis examines all the data across 10 key variable factors of our civilization. Population, fertility rates, industrial output, mortality rates, food production, finite resources, continuing pollution, human welfare, used services, and ecological footprint. All measurable and collectible data factors with which to form conclusions and make predictions, a fact that becomes important later. What was revealed was that the latest data most closely aligns our trajectory with two particular scenarios from the old study, 'BAU2', our well known Business-As-Usual, and 'CT', the lesser known Comprehensive Technology.

Both of these scenarios show a complete halt to growth, pretty much within this decade or so. That basically means that continuing to engage in business as usual and pursue continuous growth in the face of known limits is simply not possible. And even if we had incredible new technological developments that were put in place immediately, the fact remains that a business-as-usual approach inevitably leads to declines across the board in everything from industrial capacity and agricultural production to economic activity and human welfare by mid-century.

That doesn't mean civilization just ends, though. But it does indicate that a steep decline begins, and Herrington puts it at about the year 2040.

For anyone who is interested, I highly recommend reading "The Limits to Growth: A Report for the Club of Rome's Project on the Predicament of Mankind." ISBN 10: 0330241699 / ISBN 13: 9780330241694, Published by Macmillan, 1979. This will allow you to explore the original MIT study in greater detail than just what I am presenting here.

The Inevitable End of Growth

There has never been anything that is so critically important to understand as the topics in this book. And they are far from being of my own creation. I am only touching upon the life works of brilliant minds such as those MIT fellows Donella H. Meadows, Dennis L. Meadows, Jorgen Randers, William W. Behrens III, but also

Richard Heinberg, Dr Albert Bartlett, and so many others. Their contributions to our understanding are phenomenal, and there is no way I would be able to function except maybe as a data entry clerk in their presence. They are the ones who originally opened my own eyes so long ago and helped me set out on the journey to apply the findings of their work to what I could observe in the world around me.

The fundamental thing this book tries to highlight is the disparity between our global economic and social systems, which assume constant growth is possible, and the reality that we live on a finite planet. There is only so much of it, whatever "It" happens to be. It doesn't matter if the growth rate of something is 0.01% per year, if it is never allowed to go below zero (or if the long-term average is above zero), then it is fundamentally unsustainable. This is not a matter of opinion. This is simply mathematical reality.

And that is it. The root cause of all the problems we face. It is simply this: that no matter what, we will still outstrip everything the planet has to offer if we don't stabilize. It doesn't matter if we "approach" stability, or if we get close to it. We must hit that stability dead on. And that stable endpoint must be beneath the carrying capacity of whatever we are discussing; land-area, energy, resources, water, etc.

There are hard and fast limits to everything on this planet, and if we overshoot any of these limits (and we really don't know exactly where they are), then there are processes that get triggered in order to keep us within them. These things are war, disease, famine, drought, starvation, and all the other gruesome ways to die that you have heard. If we grow our population so fast that it outstrips the ability of the Earth to feed them all, for example, then a great many will starve until a balance is reached.

There is only so much growth that is possible, for anything. To understand why, look at the "infinity argument" and decide whether such a thing as infinite growth is even possible or not. For example, assuming we only have one planet, let's look at population growth. If human population keeps increasing, then one day we will have humans packed so densely that each human has only 1 square foot of space on the earth in which to live. Space is a limit, there are only so many square feet of land available. Sure, we can build skyscrapers, take advantage of growing upward when we can't grow outward but that only buys us some more time. If we still continue to keep growing, then what? Can we have skyscrapers every square foot, covering all the Earth with zero other land to use for other things like food? No. But even if we could do that, and we use the entirety of the world's resources building these skyscrapers, eventually we still hit even that limit.

So, if we know for sure that we cannot just keep growing infinitely, then there inevitably comes a point where something is guaranteed to happen that prevents it. These manifest in the real world as starvation, disease, wars over resources, all those things we mentioned. We get to where we run out of land for food and cannot feed all the people we have packed onto the planet, or we run out of energy resources to power all the daily lives of the multitudes of people, or something. And when that something gets close enough to be seen, then people start to react, and the strongest try to ensure their own survival by taking what the weaker have. Or any number of occurrences, but any way you slice it, it becomes a hard and chaotic fall from a great height. So, if we continue ignoring the reality, stop asking hard questions, and refuse to stop living unsustainable lives, then this is the endpoint.

Because it simply has to be. It is not mathematically possible otherwise.

And the thing is, we are rapidly approaching the point of no return right now. Not only have we grown massively, but the very continuance of our way of life depends on that continued growth. If we stop, which really is the only solution, then civilization collapses. But if we don't...it still collapses, just worse in scale and perhaps permanently. It is arguably one of the most interesting ethical dilemmas that humanity will ever face. It almost always leads to people attacking the person trying to bring to light our impossible lifestyles, and the terrible conclusion that the only way to avoid catastrophe is to understand that if "growth" is the problem, then the only solution is the opposite of that. Degrowth. That dirty word shared by those in the know, which means, in short, a softer, managed collapse, but a collapse all the same.

The tricky part is, this is the solution that will get exercised whether we want it to or not.

People get defensive when this type of solution is brought up. How dare we suggest that they cannot have the lives and dreams they have worked so hard for already, the ones that were always promised to them if they just did things right and followed the rules laid before them. I've had these discussions with people, and it almost always ends up in them getting angry thinking that you are attacking them. It's hard to wake people up who don't want to be woken up. It's hard to admit that the lives we live and have strived to achieve for ourselves are the things that are destroying the future for our future generations. People do not want to hear it. They just want to have their vacations and road trips, their barbeques and yacht charters, and keep their heads in the sand about the impact their lives have on all of our children's futures.

They just cannot admit that the universe doesn't care about our feelings, about our wants and desires, or our hopes and dreams. It just reacts to our actions. And if our actions are to be a parasitic species that spreads everywhere and consumes everything, then the universe will fix us with a terrible cure.

This attitude among the vast majority of the world population is why there will never be any political leaders who will suggest the necessary solutions. Any who did try and run on a platform of reducing the quality of life for people would get less of the vote than Kermit the Frog. Even between nations, no one is willing to be that first one to back away from growth for fear that they will simply be snapped up by stronger neighbors and consumed. Degrowth, as necessary as it may be, is simply a political non-starter at this point.

There is a lot of hopium out there. It is the fastest growing drug there is, because even many of those who are aware of our predicament cannot face the reality of the situation. Technological hopium is one kind. There are solutions like carbon capture technology, or grand scale geoengineering to block some of the heat of the Sun, or even the hope of spreading out across the stars to find new world to grow upon. There is also the hopium of religion, that if we just live our lives according to His Word, or These Principles, then we will all eventually be uplifted into…wherever, and there we will have eternal bliss and happiness. So, none of this life really matters, and God, well, He/They/She knows all, and would not have made things to work otherwise.

We also have those for whom hopium exists in the slim margins of what we don't yet know. We don't know exactly what the tipping point is for collapse, so there could still be time to fix it, if only we just act now…and now has been *now* for many decades of these peoples work to change things. But nothing has changed. These admirable people, they fight for change, and the try to protect the world and reduce our impact upon it, and that is such a valiant, honorable thing…but it won't win in the end. We are simply out of time, and there are too many interests that seek to maintain their Business-As-Usual to keep living the lives they want, and too many powers in the world that work to keep their power.

It is too late for the prevention of collapse. All that remains to us is preparation for it.

That is all pretty grim stuff, to be sure. But we have to look farther than even the venerable Limits to Growth study. This book here is not based on any one study alone, or even just studies in general. Because as thorough and detailed as these studies are, there is a lot that has been left out. Mostly unforeseen factors that cannot be accurately predicted due to randomness, or simply things that were unknown or unquantifiable at the time of the original study. And a great many of them are diluted or "watered down" so their authors can avoid being labeled alarmist, or just to make them more palatable to politicians. The science is there within them, but the conclusions often miss the mark, which is something else I will delve into.

The real fact is, we have actually done quite a bit worse than projected by MIT, since that study and model had no way of predicting all the unforeseen events along the way. Much of it was geared towards what we were doing to our environment in terms of resource depletion and industrial output. But, shit happens, as they say. And it seems like things have been happening at a faster than expected pace lately. There are quite a few things that must be considered alongside those well documented factors of Business-As-Usual.

There is another good study, although with a narrower focus, that was conducted by researchers from Chile and the UK. Titled "Deforestation and world population sustainability: a quantitative analysis," by Mauro Bologna & Gerardo Aquino and published in the Nature journal Scientific Reports. The study was published slightly over two years ago, in May of 2020, and it also paints a grim picture of our future.

In that study they used various modeling techniques to see how our rates of resource consumption, primarily deforestation, affects the ability of our global society to sustain itself over the long haul. The general conclusion of their work is that there's just a 10% chance that human civilization will be able to make it through the next 2 to 4 decades without suffering from a cataclysmic collapse scenario. In fact, to quote their findings, Earth's civilization has a very low probability of surviving the next few decades:

"We conclude from a statistical point of view that the probability that our civilization survives itself is less than 10% in the most optimistic scenario," they write. "Calculations show that, maintaining the actual rate of population growth and resource consumption, in particular forest consumption, we have a few decades left before an irreversible collapse of our civilization."

Granted, this study was a bit limited, but it sings along with all the others to the same general tune. And there are so many more out there, with more coming every day as things begin to accelerate. There are wide ranging studies, narrowly focused research pieces, and a plethora of very educated opinions available for just about every aspect of the coming collapse. But they all share a common theme, and that is this: whatever is happening is going to happen quite soon. Within your lifetime, dear reader. So why is does it seem like no one in power is taking it seriously?

An Existential Threat

Yet another very good paper floating about the internet has warned that society could collapse due to climate change-related disasters and conflict in the next 20 years or so, mostly because our world leaders and policy makers are not treating this as the existential risk that it is. Not a good conclusion at all. Furthermore, according to the authors, such a collapse has been deemed not just possible, but quite plausible should nations of the world fail to take meaningful action. And take action they certainly have not, nor do they show any signs of doing so.

Exactly what I have been saying here, and will continue saying, joining an ever-growing chorus of doomsayers out there who have so far been proven right.

This policy paper, titled "Existential climate-related security risk: A scenario approach," paints a scenario in which global social order breaks down after people fail to band together and address the causes and effects of climate change. And soon, like within the next twenty years. Food shortages emerge as supplies run low, financial systems buckle causing economies to collapse, sickness and disease kill people by the millions, and natural disasters ravage the land. Mass migrations of refugees from broken countries and ruined environments strain even the more resilient nations to the breaking point. Trade breaks down, nations stop co-operating with each other, and conflicts eventually break out, plunging the world into war, and possibly into darkness.

The authors, David Spratt and Ian Dunlop write, "This scenario provides a glimpse into a world of outright chaos on a path to the end of human civilization and modern society as we have known it, in which the challenges to global security are simply overwhelming and political panic becomes the norm." I highly recommend giving it a read.

They present this scenario, carefully outlined in the paper, as a potential outcome for the world in the near future should things continue as they have been, along the 'business as usual' approach. They are hoping that such a dire prediction will prompt governments around the world to treat the climate crisis as a national security issue, one that represents a clear existential threat to humanity. So far it seems like climate change has just been considered an environmental problem with health, political, and economic risks that must be offset rather than something that can bring about the collapse of global civilization, and possibly even the extinction of the human species.

This paper should not be considered as just two theorists cherry-picking data to come up with a worst-case scenario. It is anything but. The scenario outlined in the paper is actually quite plausible based on the idea that the current pattern will continue, and no one will really work together in meaningful ways to curb climate change. I would describe the presented scenario as an essential piece of information for use in scientific discussion of the collapse potential of not just climate change, but the ensuing conflict as well.

I say that not just because those many assumptions of governmental failure have proven correct over time, but because something like this can help connect the dots for other climate crisis aware people who are worried about sounding too alarmist.

It's really about expanding the conversation so that people don't get caught up in the idea of climate change just being about clean air and fresh water. Such a doomsday scenario isn't just caused by climate change alone. In reality, climate change is simply a factor, albeit a major one, that puts strain on other aspects of our geopolitical systems, economic networks, social institutions, and even parts of our increasingly intertwined cultures. As we are starting to see, it is putting stress upon the individual people as well.

Because of the complexity and interdependence of all of our interconnected markets and nations, any breakdowns can have ripple effects that result in cascading failures across the entirety of global civilization. Once you reach a certain point, major events are no longer able to be predicted with any accuracy because there are just too many factors in play, too much going on at once.

Several other scientific institutions and governments have released even more damning reports lately about the dire impact that climate change is having on such things as biodiversity, the economy, ocean health, weather patterns, and so much more. But I would argue, as did Spratt and Dunlop, that those reports are much too conservative in their estimates of what could happen. They tend to leave out a lot of negative climate change-related events that are very difficult to predict and even somewhat under-represent the risks. In fact, the authors even believe that most major climate-change reports are, just like the aforementioned IPCC report, usually edited and watered down to satisfy the leaders of so many different nations with competing political and economic agendas. And pardon me, but that's a bullshit move.

Everything seems to be about appeasement, either of politicians focusing on their re-elections, or of the shareholders of major corporations who need growth to continue forever in order to bank the profits with which they control the politicians. Even the mainstream media joins in with a narrative meant to placate the masses, and keep everyone calm, because the last thing everyone wants is a panic. They want people to be blinded by hope and faith in the system instead. However, I for one think it is high time we began to panic, and I am not alone in that opinion.

No matter how you look at it, growth has hard and fast limits, whether we choose to acknowledge them or not. And in one form or another, almost all growth is somehow driven by the fossil fuel industries. If those industries were to be stopped, so too would our society. And if they are not stopped, well, the Earth will stop them for us, with the same result. Which is why, in part, that collapse is inevitable. Degrowth is the only possible result of where we are. Either planned and implemented by us, which is highly unlikely, or a rapid and violent one dished out by the force of nature itself. Either one is a collapse scenario, the former just has a slightly softer landing than the latter.

Over time, the global climate crisis is certain to fuel competition for resources, economic distress, and social discontent through the next few years, across all societies. It will deplete our crops and people will starve. It will spawn disaster after disaster, in ever increasing strengths, all across the world. Such things will eventually lead to conflicts between nations. Potential disasters like extreme weather events, breadbasket failures, and melting sea ice are major factors that could bring countries into conflict with each other for territory and resources.

The climate crisis will be the roots of conflict around the globe and as such it poses a direct threat to the continuance of civilization. Some conflicts will become too important, like "do-or-die" situations, and the end result of such will be a nuclear war. That is always a last resort, of course, but that is what happens when things become desperate. You exercise your only remaining option.

Climate change must, therefore, be taken very seriously as a major national, and indeed global, security issue that needs to be addressed by all nations.

This paper was intended to have exactly that effect. Get the governments of the world to see climate change as a serious national security issue. But will anyone in power listen to the warnings? Can the major nations of the world actually agree on something, and then do it, before it is too late? I don't know. I hope so, but I highly doubt it. So far, it looks like we are all going to continue on as we always have. Business As Usual.

The View from the Cheap Seats

My own observational work continues with this line of thinking and expands upon the duality of climate change and conflict with yet another two c-words: Contagion and complexity. I guess we can call it my "Four C's for a 4°C World."

The first "C" is **Climate**, and within this category I include more than just the factors attributable to climate change itself, but all environmental and ecological issues that act as either accelerators or enhancers of the climate crisis itself. Pollution, resource depletion, the loss of biodiversity and even the scarcity of energy, all of those are individual force-multipliers of climate change.

Next up is my second "C" of **Conflict**. Other than just the new world war we seem to be currently getting ramped up, there are also issues of societal and economic conflict that are presented within this category, as well as future wars that are just now in the budding stage, as well as the inevitable outbreak of the global resource wars. There are wars going on within us as individuals as well, and then there is the all-encompassing war on truth by the fossil fuel industry.

After conflict comes the "C" of **Contagion**. From COVID-19 and the newly emerging monkeypox, to bird flu and antibiotic resistant bacteria, and even to the new problem of ancient pathogens lying in wait inside the quickly thawing ice of glaciers and permafrost, contagious diseases are a scary independent enemy all their own. Some are here among us already, but others are waiting in the wings.

Finally, there is **Complexity**, my fourth and final "C.". This deals with the problems of our incredibly complex world of interdependent systems, with every aspect of civilization wrapped up in one big machine just waiting for the right wrench to be thrown into the works. We explore how everything is so vastly complicated in our world that we have abdicated almost all function to automated systems and algorithms not under any real human control.

I have included as many different things as I can, given the limits of my own knowledge and expertise, but the list of dangerous crises I present here is by no means complete. However, this should serve as a nice base for you to see not only how many things have gone wrong, but also how everything plays off everything else. I hope this work can be something of a springboard that can help launch your own expanded understanding of the situation.

When reading the material in these various categories, keep in mind the rules we have already established about those hard and fast limits to growth, as they play a part in all of them. With that, let's get right into it. As you should know by now, there is not much time left to waste.

3

THE CRISIS OF CLIMATE CHANGE

The First C: Climate

The understanding and documenting of climate change was still in its infancy back in the 1970s, at least compared to the horror story of what we know today. We not only know much more about the crisis now, but more importantly we know just how much we really don't know, something that is becoming clearer every day as the effects begin to pile on.

Just recently, the Intergovernmental Panel on Climate Change released its sixth assessment report in full. The IPCC is the scientific group assembled by the United Nations to monitor and assess all global science related to climate change. It was created to provide the world's leaders and policymakers with regular scientific assessments on climate change, the implications and potential future risks it holds for us, as well as to put forth adaptation and mitigation options for consideration in forming new policies. Through its assessments, the IPCC's job is to determine and codify the state of human knowledge on the climate crisis. Then, the findings are presented to world leaders, along with recommendations for what to do about it. However, it is deeply flawed in many ways, some of them intentional, but I will get into that later.

This recent report was a dire warning indeed. In fact, they used the words "A Red Alert for Humanity" as a description at some point. There is no need to break down the report here, and in fact such would not even be possible. Go check it out yourself, it makes for a very harrowing read. Instead, we can stick to the salient facts and go with the conclusion in summary.

I am not focusing very much on the past reports, as really, they are all just a succession of the same warnings that we have been steadily ignoring for a long time. The difference with this particular report is that it may well be the last one the IPCC puts out with even a slim chance of mitigation of the effects of climate change. "It's now or never," were the exact words used by the presenters, I believe. The final report is a chilling assessment of where things stand. Without a dramatic and rapid course correction we are heading for over 3.5 degrees Celsius of warming by the end of the century, at best. This is a catastrophic level of increase that scientists say would drown some coastal cities, intensify heat waves, droughts and floods, and make swaths of the world all but uninhabitable, driving starvation, species extinction, disease, conflict and mass migration, just to name a few of the effects. And the reality is that warming well above 3.5°C is most likely already "baked in", if you will forgive the pun.

According to the report, all greenhouse gas emissions must peak by no later than 2025 in order for there to be even a chance to limit global warming close to 1.5 degrees Celsius as targeted by the Paris Agreement. That is just two-and-a-half years from right now as I write this. Furthermore, those emissions must then be cut in half from that peak by 2030. And that is only to have a chance to limit warming close to 1.5 degrees Celsius. But even with "only" 1.5 degrees Celsius of warming, the world faces a great acceleration and intensification of climate impacts like killer hurricanes and wildfires, crop-decimating heatwaves, and so on. And, without immediate and drastic emissions reductions across all sectors, it will be impossible to reach even that rosy result. 1.5 degrees is basically a pipe dream at this point.

There is a word that is used to describe what happens if nations of the world fail to meet the challenge of stopping our destructive path. That word is "overshoot." It might not be four letters, but this is indeed a bad word.

Overshoot occurs when the world's populations and their consumption grows rapidly to the point where the demand on the planets ecosystem exceeds the capacity of that ecosystem to regenerate the resources that are consumed, and to absorb the waste produced. Basically, this occurs when man's demand on the biosphere exceeds the available biological capacity of the planet. Fossil fuels and the industrial revolution allowed us to grow too much, too fast.

We were like a rocket moving upwards on the charts. And along the way, even the rates of our consumption per capita increased incredibly. Somewhere along that upward trajectory, there was a line which indicated the max carrying capacity of our planet, but we were moving up much too fast. We blew right past that line, overshot it completely, and the momentum of the drive for infinite growth kept us going up and up. But the problem with overshoot is that, just like with a rocket, at some point the fuel runs out. We may coast for a time, still upwards, but a moment comes when we hit the peak of our ballistic arc. And then, unable to sustain the level of height we have reached, suddenly we tip over the arc and begin a dramatic and accelerating plunge downward. During that dive, we build new momentum, only downward this time, and just like before we blow right through that line indicating our carrying capacity. Now, we are rocketing just as fast as before, but headed down. During the overshoot we depleted the planet's life supporting biological capital and accumulated more waste products than it could absorb. This means a permanent reduction in total carrying capacity for the future, even after we hit our new rock bottom.

Overshoot is collapse. And right now, all these things we are starting to feel and experience? That's the peak. We are there, rapidly losing momentum and starting the final drift before the plunge. What is coming very soon is the new rapid acceleration downwards, and how we deal with that will determine just how hard our crashlanding at the bottom will be. But either way, hard or soft landing, we are going down.

Something amazing to me is that there is even an idea going around out there that we can "make up" for overshoot if (when) it happens, and that is dangerous thinking. Because the main ingredient of this idea is that we will be able to use nature or not-yet-available technology to draw down our greenhouse-gas levels later, and by doing so we can return the temperature back below the limit. Or replace our current overconsumption of resources with new energy and material sources by the time the other stuff runs out. So, basically the plan would be to just let everything break, even though we don't know how to fix it, and trust the idea that we will have figured out how to fix it by the time it actually breaks down completely. Yep, really, that's the idea. But the new IPCC report warns even if we were somehow able to do that, a really big if, there will still be many more severe effects on the world, many of which are irreversible compared to just avoiding overshoot to begin with.

Simply put, an increase of more than 1.5°C, basically a given at this point, would thoroughly undercut the possibilities for many kinds of successful adaptation to the changes, and open up an entirely new frontier of danger for every fraction of a degree we shoot beyond that temperature. The report stated its findings that it was now "almost inevitable" that temperatures would rise above 1.5°C. This is the level that, if we go over it, most of the effects of climate breakdown will become irreversible. The effects will begin snowballing, getting bigger and moving faster, until we are completely unable to cope with them. We can see it happening now. Ever since those massive wildfires in Australia for 2019, it has been one hit after another, and we are getting stretched to the breaking point. As bad as it has been these last few years, think about this: that was with our systems actually working to try and recover. We are running around right now, collectively sticking our fingers in holes to stop the leaks, but what happens when we run out of fingers? What do you think it will be like when the dam finally bursts?

To avoid a disastrous series of cascading and compounding damages to critical human and natural systems, the world must move away from its current sluggish, and virtually nonexistent adaptation measures. There can be none of the usual instances of nations abstaining from some parts of the plan or rolling back decisions after a year for political reasons, and indeed there can be no independence of action at all. Every nation, every single one, must immediately move to a sweeping program of anticipatory, resilient planning and development for the reduction of fossil fuel use across every sector if industry it touches. In fact, what is in the ground now, it has to *stay in the ground*. This would require intense cooperation across every sector of global society and industry, and at every level of governance. And it has to happen very quickly, before the world exceeds those temperature limits. Short of actually collapsing, there really is no way such things could be done quickly enough.

In a way, this report is kind of a final warning for us, because at the IPCC's pace of doing new evaluations every 7 years or so, the world almost certainly will have burned through its carbon budget by the time the panel releases its next climate mitigation report. Now or never, indeed. By the time these people get together again, there may well be no 'mitigation' section to the report at all, just a cold and scientific cataloging of how bad we have made things and what we can try and do to adapt and survive.

Look around right now. This warning came just a few short months ago. A desperate pleading from the scientific community for us to all work together to tackle this problem, to take advantage of the, and I quote here, "brief and rapidly closing window of opportunity to secure a livable and sustainable future for all."

Brief and rapidly closing window to a livable future. What did we do with those words? Threw them right in the trash, that's what we did. Right now, nations around the world are expanding new oil and gas infrastructure as fast as they can in the face of the new war breaking out in Europe. Many of these countries are even firing up their coal burning power plants again, and I am not just talking about long-time contrary nations like Vietnam, I am talking about Germany firing up the coal burners. Coal. The dirtiest fossil fuel of all. Austria, the Netherlands, Italy…all of them and more. They didn't even wait more than a couple months to backtrack on their climate pledges in favor of firing up coal power plants. That doesn't give me much hope for them doing anything other digging our hole deeper while our last window of opportunity closes for good.

And it gets worse. Because, as damning as the IPCC report was, it was only a watered-down version of the true potential. This thing actually paints the best-case scenario, and to do that relies on an enormous amount of faith and techno-hopium. The third section of the IPCC report is the mitigation part, which outlines things society can do to slow down the trajectory of climate change. The problem is that these researchers are forced to reconcile the scientific realities of the rest of the report with economic and political considerations that are not based in science. Many people have described this part of the report as a mechanism to determine what is politically possible, rather than what is actually needed. This is one of those big flaws I was talking about when it comes to these reports.

Due to that, they depend heavily on assumptions of things that are currently not within the realm of reality. If those assumed solutions don't materialize, such as the future availability of carbon dioxide removal technology which is currently beyond what is possible, then all you really have left are illusions of hope. Besides, even if the technology was there, the simple fact is that if it is not profitable to do so no one is going to invest the required trillions of dollars needed to accomplish any of it on a global stage.

What they are doing is trying to put forth solutions that still allow humanity to continue with its constant search for that infinite growth we hold so dear, while at the same time fighting climate change, and that is simply wishful thinking. But politicians and policy makers do not what to hear such things, because stalling or reversing growth is just not politically possible. In fact, it would be career suicide for any world leader to suggest any such thing.

The only real 'carbon capture' at work these days are the great and powerful fossil fuel corporate interests which have captured the loyalty of the governments of the world and turned our leaders in fossil fools.

There is only so much more carbon dioxide that can be emitted before the world misses the Paris Agreement target. That is our 'carbon budget.' There are only about 400 gigatons left in that budget, and this new report shows that civilization is on a path to produce more than twice as much, and that is just taking into account the building and use of things that are already currently planned and scheduled. Not to mention the planned "carbon bombs," which I will get into later on. We are literally doubling down on fossil fuels.

And now that the war between Russia and Ukraine has kicked off, the leaders of the world have been talking about expanding these programs for even more fossil fuel infrastructure. As these nations are firing up the coal power plants again, they are also starting new drilling and exploration for oil and gas as well.

The IPCC report, in its entirety, is incredibly grim, even in its watered-down format. But you won't hear the mainstream media addressing that issue with any real gravity. And that in itself is just a continuation of the goal to downplay and sidestep the issue in the interests of continued economic growth. No matter what happens, the media or some "expert" will always throw out hopium of some sort to the masses. Somehow, we can spend, buy, grow, or work our way out of this mess, according to them. That is the message sent to give you hope, but the purpose behind it is not to actually solve anything. It is just to keep the people moving and going to work and having faith that we will all pull up our bootstraps and fix it all so we can "get back to normal."

If you haven't seen the movie "Don't Look up" I encourage you to do so. It may be a comedy, but it is also a documentary in many ways as well.

Environmental Collapse

Our global ecosystem is the foundation for all life on Earth, and that especially includes us humans. Ecological processes and systems within our environment perform a wide range of functions, and without them human civilization could not operate at the current level or anything close to it. We depend on the environment to provide our air, water, food, shelter and energy. Everything we need to continue to stay alive and build the complex civilization that we have, it is all dependent on the health of the environment.

This environment can tolerate a decent amount of impact and abuse from human activity and still recover after a while with little in the way of permanent negative effects. This is often referred to as the attribute of resilience. The Earth, while being a fragile system, is still very resilient and capable of absorbing and repairing an incredible amount of damage.

But what happens when we push things too hard, try to grow too quickly or use up resources faster than they can be renewed? When we constantly despoil and pollute the environment, use up non-renewable resources at a fast pace, and work to destroy the ecological systems that our very survival depends upon? Beyond a certain threshold, called a "tipping point", a rapid and irreversible disruption occurs, which leads to "ecosystem collapse." The beginning of such a collapse can be seen and felt by a sudden accelerating rate of change in our world. Under collapse conditions, things like soil quality, freshwater supply and biodiversity begin diminishing drastically. Our oceans produce less fish, and our weather patterns become more chaotic and severe disasters occur more frequently. Agricultural capacity begins to plummet, and many aspects of our daily human living conditions deteriorate significantly.

Now, ask yourself, doesn't that seem exactly like how things are going right now? You may have already felt some of it personally, and certainly you have seen it in the reports from across the world.

Although not as widely reported as it should be, new evidence has been emerging on climate change and ecosystem collapse. And increasingly we have begun to see the grudging admission that much of the problem is due to human pressures and expansion, energy use, and resource depletion. Where once the subject of climate change was a little murkier and always distant, it has become too noticeable to ignore for most, and is even making the most hardheaded of skeptics take a closer look at their own denial.

Local ecological collapse has caused the end of civilizations before. Just look at Easter Island. More recently, ecological collapse in and around the Aral Sea has had dramatic social and economic consequences for the region, although timely intervention has led to some small recovery as a temporary measure. Ecological changes in around the world have not only destroyed livelihoods with dramatic impact on people and ecosystems, but the increasing climate disasters and diminishing resource availability has led to severe health and economic impacts for the people of almost every region. We are even starting to see the generation of resource-based conflicts.

In the highly connected world of today, local disruptions can also lead to unintended effects in many different far-flung areas. This will eventually escalate into the rapid collapse of most ecosystems across the entire world. And with no time left, or even political will, for effective recovery, we are dooming ourselves to experience the full force of ecological collapse in our lifetimes. Even now, in these early stages, everything we do has its difficulty amplified by various climate change impacts, whether it is growing our food, powering our cities, or moving our goods where they need to go. There is no other way to look at it, the simple fact is that we have drastically compromised the planet's capacity to sustainably support our human population. We have, at every turn, sought growth over conservation, and the price of that action is coming due. In short, we have fucked around, and now we are about to find out.

Researchers and scientists have come to describe our current moment in the Earth's history as the start of an entirely new geological era, called the Anthropocene. Within this new era humans have become the predominant agent of change at the planetary level and have even changed the nature of the environment itself. Since the dawn of the industrial revolution, most of the factors that ensure the habitability of the planet have been degrading at an accelerating pace. Whether it is greenhouse gas concentration in the atmosphere, the reduction in the health of forested areas or marine ecosystems, or a variety of other factors, humans have been forcing this planet to its limits, and beyond.

Back in 2009, an international group of experts identified these limits for us. Johan Rockstrom who is the founding director of Sweden's Stockholm Resilience Centre, gathered together a large team of very accomplished scientists to unite behind the goal to define the boundaries for a "safe operating space for humanity" on Earth. They classified a total of nine interconnected, and interdependent, planetary boundaries that form the foundation of stability which the global ecosystem rests upon, and that foundation is what allows human civilization to exist. It has been argued that we have already exceeded the safe limits for several of these planetary boundaries and are now operating in a high-risk zone for maintaining biosphere integrity. Furthermore, given our current pace and activity, we are very likely to exceed all nine of those boundaries soon, and move beyond the safe operating space where humanity has thrived, and violently overshoot our planet's carrying capacity for life.

These are the nine planetary boundaries beyond which we can't push Earth's systems without putting it at risk of complete ecological collapse:

1) Climate change
2) Biodiversity loss
3) Ocean acidification
4) Ozone depletion
5) Atmospheric aerosol pollution
6) Freshwater use
7) Biogeochemical flows of nitrogen and phosphorus
8) Land-system change
9) Release of novel entities

Human civilization is already existing far outside the safe operating space for at least five of those nine boundaries. Climate change, biodiversity loss, land-system change, release of novel entities (meaning pollutants such as plastics), and biogeochemical flows (meaning a nitrogen and phosphorus imbalance). At this point in time, there really isn't much hope that we will prevent catastrophic overshoot of at least these five. I highly recommend checking out the data for yourself at www.stockholmresilience.org

New evidence suggests that the time remaining to change course and stop the accelerating human-driven decline of life on Earth is quickly running out. This is no longer a set of consequences that we are looking at facing many generations in the future, they are going to be felt now, this generation. In fact, given the state of society and the political realities our leaders and corporations are beholden to, such a reversal of decline is most likely impossible now, and even some measure of mitigation or adaptation to what is coming requires the same kind of radical and transformative change that we failed to enact to prevent this in the first place. Now, even those chances of mitigation are fading fast.

The Many Facets of Climate Change

Many people have the general idea that climate change just means warmer temperatures around the globe. But increasing temperature is only the beginning of a very long story for the climate crisis. The Earth exists as a very large and complex system. Everything is connected to everything else, and all of it is interdependent. Large changes in one area can influence outcomes in many others. Our planet is such a complex system that we can barely even grasp the surface of what climate change can really mean.

But many of the consequences of climate change are becoming clearer, especially as they begin to manifest around us. We now know that some of those effects include intense droughts and water scarcity, severe and unpredictable wildfires, rising sea levels and catastrophic flooding from melting polar ice, dramatic reductions in biodiversity from mass extinctions, and storms of an increasingly unprecedented ferocity striking in unpredictable locations.

People everywhere are experiencing climate change in many different ways. It is affecting our health, our ability to grow enough food to feed our growing population, and our physical safety at home and at work. We are seeing the increase in energy use to combat temperature swings, and the rising costs of repairing the damage done by the violence of nature. Some people are even more vulnerable to climate impacts, such as those living along the coastal areas or in small island nations. factors such as sea-level rise and saltwater intrusion have progressed to the point where whole communities have had to relocate, and the same can be said of the raging wildfires that have swept aside entire towns in some places. In the future, the number of "climate refugees" migrating away from fragile areas or the tropics and desert lands is expected to rise considerably, and that will be a crisis in itself, putting enormous strain on other nations while also increasing the instances of civil conflict and unrest.

The more the planet heats up, the greater the damage to the climate system becomes. Things begin to get more unstable, and that instability feeds on itself, increasing more and more until civilization can simply no longer cope with the damage. Things that we all experience from time to time become worse with each repetition. This includes more lasting and intense hot extremes, ocean temperature heatwaves, heavy precipitation which dumps more moisture in shorter periods, agricultural and ecological collapses due to extended droughts, the proportion of intense tropical storms and hurricanes, and reductions in Arctic Sea ice which will eventually culminate in a Blue Ocean Event. More on that in a moment.

Continued global warming will further intensify the global water cycle, making it more variable, and changing monsoon precipitation and the severity of wet and dry events. Agriculture depends on predictable weather patterns, so how will we feed ourselves as the weather becomes increasingly chaotic? As greenhouse gas emissions rise, the ocean and land will be less effective at absorbing them, which will only speed up their accumulation in the atmosphere. Deforestation on land and increased pollution of the oceans is driving this even faster. Such damaged ecosystems tend to emit more carbon instead of storing it, and the Amazon rainforest is one such area which used to be a carbon sink but has now become a carbon emitter.

There are a great many ways in which the climate crisis will lead to our collapse. So many so, that I would never be able to go into detail about all of them here. But we shall explore a few of the more serious ones, just to get a feel for how truly screwed our civilization is, and just how close to collapse we really are. I am also going to rope some other environmental factors into this category, things that are not directly caused by climate change, but that are either exacerbated by it or contributing to it. Things related to those 9 planetary boundaries we have already gone over. All things ecological and environmental are going under this "climate" umbrella, but keep in mind that as bad as the things I am writing about here are, this is by no means an exhaustive list. I'm just going to hit some of the majors, so let's get to it and work our way through this horror story one step at a time.

Blue Ocean Event (BOE)

The coming Blue Ocean Event is probably something you've never heard of, since for the most part the media tries to ignore the really bad stuff, especially when it is a guaranteed one. But in a few years' time, at best, it will be a huge story that everyone will be talking about. This Blue Ocean Event is when the Arctic Ocean undergoes a change of state from being covered in ice all year-round, which reflects most of the sunlight that hits it back into space. In its new state it will be mostly ice-free blue water for a long period of time during the warm season, which will cause it to absorb much more of the sunlight that shines upon it.

Such a thing has never occurred since our ancient ancestors started wandering around the Earth millions of years ago. So, never in the history, recorded or not, of the human species. This will have an incredibly negative impact on the Earth's climate and natural habitats, and all of civilization. Once one of these events occur during a warm season, it won't remain so. Not immediately, anyway. The water of the Arctic Ocean water will refreeze during the next cold season. However, since that water will be getting slightly warmer every year, it is expected that the Blue Ocean Events will last longer each time, causing extensive impacts to the climate of our world.

Why does such an annual melting of arctic sea ice matter? There are several reasons, but it is primarily because once a BOE starts occurring every year, the Earth's atmosphere will start to warm up even faster than it already is. We can all see how global warming has been progressing, and speeding up, but this new effect will cause much more rapid warming than ever before, and that will have major impacts on the climate that civilization relies upon for stability to grow the crops that feed the world. Less food, less humans. And that transition to 'less humans' won't be a peaceful one.

Even the slow, gradual climate change that we are just now starting to admit has already begun affecting the food supply. Droughts are becoming more severe and longer lasting, for instance, and along with unpredictable and out of season frosts and heatwaves, there has already been significant crop loss in recent years. There is a lot more, but you get the idea.

Here is a quick breakdown of how a BOE works. Blue water, which simply means darker water, absorbs quite a bit more sunlight than white ice does. In the phenomenon known as the "albedo effect" darker colors are naturally prone to absorb more sunlight than lighter colors do. The additional solar radiation being absorbed by all that newly blue water will heat up the ocean region even faster than temperatures are already warming, which will cause less ice to form during the following freeze and even more ice to melt during the warm season after that. This will just begin a cycle of extending the length of each BOE year after year, until one day there is no freeze at all...

This cycle of lengthening BOEs that causes additional global warming is known as a positive feedback loop. That is a reoccurring and self-perpetuating process that over time increases its own effects. Meaning basically that once it happens, the more it will happen, and the more it happens the harder it will continue to happen indefinitely. Scientists across the world have predicted that this BOE feedback loop will cause the Arctic region to become so warm that it will transition to being ice-free year-round at some point before the turn of the century. And they say 'at some point' because they don't really know, exactly. That's what happens when you are talking about unprecedented events. There is no real data to draw upon for the purposes of prediction.

What does all this mean for humanity? Well, most people don't live near the Arctic Ocean, right? We live mainly in areas thousands of miles away from the polar region. For many, melting artic sea ice seems like a very far away problem. The truth is, BOEs might happen far away, but despite the distance, these events will have significant catastrophic impacts on populations all over the world. Impacts that will certainly affect life-sustaining water and food supplies globally in a very negative way.

The jet stream is a current of fast-moving air in the atmosphere, like a river in a way, and it steers all the weather effects around the mid-latitudes. This weather has been stable and generally predictable in providing the rain that is used to grow crops and replenish water supplies. So stable that it is the source for our agricultural growing zones and such. We generally know what the weather will be like any given time of year in all the regions of the world, and agriculture requires such stability.

The jet stream obtains its energy and positioning in the world as a result of the difference in temperature between the cold regions at the poles and the warm region at the Equator. Once these BOEs begin persisting for many months and the Arctic grows considerably warmer, the jet stream will begin to lose its energy due to the temperature difference between these regions. At that point, storms that form in the atmosphere won't move according to their old patterns and could become stuck in place or just drift along very slowly. They won't move as much as they will have little to no jet stream currents to push them along.

This will cause excessive rainfall in areas which do get storms, dumping so much rain that crops cannot be sustained in the face of it, not to mention the flooding effects on human settlements. In other places, long-lasting heat waves and droughts will occur because of that same lack of jet stream propulsion. This will allow the creation of massive rain-free heat domes that stay in place for long periods. Obviously, that would be devastating for a civilization that relies upon steady sources of food and water. Crops simply cannot grow during extended heat waves and droughts, and our food won't grow in flooded fields either.

Essentially, a reoccurring Blue Ocean Event would be catastrophic, and indeed the end of the world as we know it. Civilization has been built on the ability to grow crops reliably and get the food to people who live far away from the source in order to sustain our societies. Civilization has also been built on reliable access to clean drinking water. Think about what happens if both of those things are no longer available. Due to a BOE weakening the jet stream and the accelerated pace of global warming, people will have to rapidly adapt to survive.

And the sad fact is that most will not be able to. Many people will die in this scenario, as large portions of crop-producing regions transition to become arid or desert climates. Global famine and a lack of access to drinking water will impact billions of people, causing mass casualties among the population. And that won't be all it causes, as such food scarcity and skyrocketing prices will lead to both civil unrest as well as conflicts between nations. Those people that survive will only do so in smaller numbers that are more sustainable in this new collapsed world, and they will have to quickly find ways to produce food and obtain drinking water in the new climate reality.

This is not one of those 'if' things. As far as a Blue Ocean Event goes, it is just a matter of time. Could even happen this year, although most predictions I have read seem to think 2025 is a reasonable target. If things continue as they have. Unfortunately, Business As usual is one things we humans are very good at.

Multiple Breadbasket Failure

In a smooth, and related, transition for Blue Ocean Events, let us now take a look at what is known as a Multiple Breadbasket Failure, or MBF. I'm not sure if that is an actual acronym, but I see no reason to type any more than I have to, since I am not very good at it, so MBF it is.

A 'breadbasket' is a major cereal-producing region of our world, and the Earth has six primary ones. They are as follows:

1) The Canadian prairies and US Midwest
2) Brazil and Argentina
3) Northwestern Europe
4) Southern Russia and Ukraine
5) Eastern China
6) Northern India

A multiple breadbasket failure is when breadbaskets in several of these areas collapse all at once, failing to produce enough food for the growing population of our world. Because so many people rely on the staple foods breadbaskets produce, such as wheat, rice, and corn, an MBF could quickly lead to a global famine of epic proportions.

One of the greatest risks we face over the next few decades, especially in light of accelerating climate change, would be coinciding extreme weather events that would negatively impact agriculture in the breadbasket regions of the world. Extended droughts, major heat waves, catastrophic floods, and raging wildfires are all things that are already starting to happen around the world, with significant damage and destruction to crops. The increasing nature of these events, in both frequency and severity, will significantly reduce food production and distribution for all the world's people.

In times of regular climate conditions, the global food system is pretty resilient and occasional crop losses and damage can be compensated for through storage and trade. Crop failures due to extreme events in one of our breadbasket regions can usually be taken care of by food production surpluses in the other areas. But that system will not be resilient enough for a situation where multiple, or all, breadbaskets are being hit by extreme events at the same time. Our agricultural system, like most human systems, is designed to function properly only if not disrupted. There is very little in the way of 'extra' capacity or room for error, because to do such would be a drag on the profit that could be gained from the system. So, when major unexpected events occur, the result is enough disruption to cause a complete failure. A good example is the war currently going on between Russia and Ukraine, which just happens to be smack in the middle of one of the world's largest breadbaskets.

If there is even a single breadbasket failure, such as the above example, it results in an immediate shock to the entire global food system. As nations scramble to find other sources for their import needs, the surplus is gobbled up relatively quickly, and speculation in the commodities markets can drive prices even higher when combined with the shortage. Those higher prices that result often trigger political unrest, conflict and migration. And that is just a single 'shock' to the global food system. What do you think would happen in the case of multiple shocks, or a complete failure?

Over the past few decades, many of the world's primary breadbaskets have experienced shocks that caused major, rapid drops in food production. For example, regional droughts and heat waves in the Ukraine and Russia have damaged wheat crops and caused global wheat prices to spike significantly a couple times in recent years. Back in 2012 heat and drought in America cut the national corn, soybean and other crop yields by something like 27%. And production yields of vital food crops are low and getting lower by the season in many other countries due to things like plant diseases, poor soil quality and topsoil erosion, poor management of agricultural assets and increasing damage from the pollution of the water, ground, and air.

On top of all that, many experts in the field are saying that we may have to double global food production by 2050 to meet the needs of a growing population and satisfy fast-rising demand for meat, and dairy products among the population in developing nations. Global food production has managed to rise over the past 50 years, largely spurred on by improvements in agricultural practices, advances in genetically modified plants, and more intensive use of things like mechanized equipment, fertilizers and pesticides. There has also been continued drive to bring new land into production. This has relieved pressure somewhat, but there is a cost.

As we all know but rarely acknowledge, growth has its limits, and the need to produce more food has been one of the main factors that are pushing these limits in recent decades. Deforestation is a serious issue that is being rapidly accelerated due to the need for more agricultural land. Large scale agriculture contributes more towards climate change than the media ever talks about. Agriculture as a whole is a major net emitter of greenhouse gases, especially animal agriculture. The needs for irrigation and water sources for all this farmland does nothing but grow while the supply of that resource has done nothing but shrink. Pollution from agricultural runoff, loss of biodiversity from heavy pesticides and tilling over the wild lands, topsoil erosion threatening future crops… The list goes on and on. The limits are real, and we are already shooting past them.

There are many different factors that make the global food system more sensitive to the effects of climate change. There is the general dependence of the human diet on a only a handful of different grains, mainly rice, wheat, corn, and soy. They make up about half of the caloric intake in an average person's diet. Such dependence on so little variety is a big vulnerability. There is also the fact of the geographic concentration of production, having most of the world's food produced in only the six major breadbaskets. This means extreme weather events, or human events like wars, in those regions could affect a large portion of global food production. You could say all of our eggs are in very few baskets, so to speak.

Another big factor is the growing dependency on grain imports by many nations, especially ones with large populations. The number of people worldwide that rely most on importing grains is growing. In particular, those countries we refer to as "developing" tend to be major net importers of grain, mostly because artificially created competitive practices in growing grains make buying from the world markets much cheaper and easier than trying to produce what is needed domestically. These nations also tend to be more susceptible to uprisings and civil unrest, so suddenly waking up one day and not having anything to eat will not be received well by the locals.

There is also the issue of limited grain storage in many nations. The amount of grain that can be stored by a nation influences how well its food system is able to deal with any shortage of food production. Such capacity provides a buffer that can be built up in years with surplus production and released in years that have shortages. But, for many smaller high population nations, there is little in the way of sufficient storage capacity. And despite historically high levels of global capacity today, grain storage levels are insufficient to withstand a large shock in production. It must also be considered how, in times of trouble, producing nations will stop most exports in order to secure food supplies for their own populations, like what India has done this year during the Ukraine war combined with crop losses from heatwaves and drought. Or how China recently started stockpiling grain ahead of time in anticipation of future troubles. We humans tend to share in times of plenty, but hard times bring out the hardness in people, and one must look out for one's own.

Most analysis done lately suggests that a multiple-breadbasket failure, one with simultaneous shocks to crops due to severe climate events in more than one of our breadbaskets, becomes very likely in the decades ahead. Driven by an increase in both the likelihood and the severity of climate events, the continued pollution of the environment, and the rising possibility of another world war, among other things.

Hunger and malnutrition are already serious issues in our world. They have been rising steadily for a while. And that's with the system actually working as intended. Indeed, even a 20% shock to the system could kill hundreds of millions, if not billions, of people and the resulting wars would kill millions more and exacerbate the food problem for the next wave, if not cause a full collapse of the system entirely.

Biodiversity Loss: The 6th Mass Extinction Event

Humans alive today are actually witnessing the early beginning of the Earth's first mass extinction event in over 65 million years. It is underway right now. We don't have to guess when it will hit or what the earliest effects will be to the ecosystem and environment, because it is already happening, and we can watch it live.

It's been about 65 million years since the last mass extinction, which was the Earth's 5th. This was the one that marked the end of dinosaurs roaming the planet, scientists are warning that we are in the early days of another such event. Not annihilation by a giant space rock this time, though. Unlike any of the others, this sixth mass die-off, also called the Anthropocene extinction, is directly caused by humans. The dinosaurs managed to go millions of years without screwing themselves over, and even in their end they remained innocent of it. Not us. In a relatively short period of time, humanity has managed to wipe out more species than I care to count right now, and we are not done yet. Climate change, habitat destruction, pollution and industrial agriculture all play a role in this human sponsored game, and in the end, there really are no winners.

In defining mass extinctions, it generally means that at least 75% of all species cease to exist within a span of about 3 million years. Given how little time we have had, it is truly amazing to think about how much damage we have done. Many scientists and researchers now believe that at our current rate, we are essentially right on track to lose that number of species within a few centuries. Just over the next few decades alone, at least 1 million unique species are at risk of being wiped out. That's according to the estimate of a landmark report published back in 2019, but there are a lot of scientists now saying it could well be an overly conservative count. Trying to predict what could happen from a complete collapse in biodiversity is almost like some kind of magical divination. The Earth's ecosystems are incredibly complex, intertwined, and interdependent.

Almost all scientists do agree, however, that there are several clear results predicted should extinctions continue to be logged at this rate. And all the effects are inextricably linked, like a giant web.

Biodiversity Loss Affects Us All

Humans rely on well-functioning and diverse ecosystems for many reasons. They provide us with clean air, fresh water, and food security. They also limit the spread of diseases and help to stabilize and regulate the global climate. But with biodiversity loss happening at unprecedented rates, the impacts to humanity worldwide will be a heavy one.

In 2020, the global public was abruptly reminded that pandemics can pose a serious threat to public health and economies on worldwide scales. With biodiversity disappearing at an alarming rate, and humans encroaching on ever more dwindling animal habitats, we are seeing infectious diseases increasingly spill over from wildlife to humans. Dwindling biodiversity will cause many more humans to contract infectious diseases that "spillover" from animals. COVID-19, probably from bats, was an example of this that truly rocked the world.

In addition to the increased risk of pandemics, one of the really big things we will see is that our food supply begins to become increasingly unstable and starts to dwindle down in terms of crop yield. Because so much of our food really depends on pollination by insects and birds. In fact, about 30% or so of civilizations global food supply relies on pollinators such as bees. If they die out, agricultural yields could go into a steep decline.

The quality of the soil we depend on will also deteriorate as critical microorganisms and insects die off. Such loss of biodiversity is severely underrepresented in the data, but some studies do show that they are quite probably vanishing at a faster rate than other types of animal life. Their disappearance will lead to a worsening of soil erosion, which then results in more floods, as well as poorer fertility for plant life. Yet another big impact to crop growth.

There are also millions of people that rely on hunting wild animals, both for nutrition and for their livelihoods. This impact will particularly affect people in coastal and island communities, which are especially vulnerable to disappearing sea life.

Ocean Health

And speaking of sea life, ocean biodiversity loss is much more serious than most people realize. The ocean is home to millions of species, all of them playing their own unique role in maintaining the balance of their environment. The health of the Earth's oceans is highly dependent upon this diversity of marine life. Biodiversity in ocean life is an essential component of climate regulation. Climate change and pollution due to human activity has a direct negative impact on marine species. It alters almost every aspect of their existence, such as the abundance of them, their diversity, and also how they are distributed throughout the seas. How they feed can be changed, their development in various life stages and breeding characteristics also, as well as the different relationships between the species. All of these and more are affected.

Rising ocean acidification, caused by the increasing absorption of atmospheric carbon dioxide (CO_2), has a phenomenal negative impact on the marine organisms with shells: these include phytoplankton, crustaceans, mollusks, and the rest.

Due to the many different changes that have been accumulating in marine ecosystems, the ocean is becoming dangerously sensitive to the effects of climate change. In fact, the more the oceans become depleted, the less they are able to adapt to climate change, and also the less they can help regulate it. The incredible pressures we have put on our oceans and on marine life is just another one of those tipping points which accelerate everything to the point of collapse. The severity of this situation has been downplayed and ignored, barely even considered by policy makers.

All of us humans rely on the ocean, whether we really know it or not. Our planet's vast expanses of water are the key to success for all life on Earth, not just us people. We get most of our fish from the ocean to eat, we need the oxygen it gives off to breathe, and the warmth of its huge currents help regulate our weather. Without healthy oceans, humans simply cannot survive.

But the ocean has taken a beating from us not treating it well. It is being stretched to a breaking point that is rapidly approaching. It is quite possible that we have already passed that point, but certainly if we don't stop depleting it further the ocean will be drastically changed very soon. Something we will see in our lifetime.

Torrents of pollution flow into the ocean every day, with no real sign of slowing down at all. A large part of it is plastic, which fish, birds, and sea mammals now regularly eat. Many of them die from it, whether from eating it or becoming entangled in it. It has been estimated that by mid-century there may be more plastic in the sea than there is fish. As all of this plastic piles up, fish slowly disappear. Since major commercial fishing by humans began, the oceans have been transformed. Today's seas contain only 10% of the marlin, tuna, sharks and other large fish that were found in the 1950s.

Overfishing knocks the entire marine ecosystem out of balance, and that has consequences which will affect us all. If we had fished more sustainably over the years, the picture would be much brighter today. If we had all just ate and wasted seafood less regularly, allowing stocks time to recover, fish populations might have been able to keep up with human appetites. Unfortunately, by this time, given how we still continue with even more industrialized fishing practices on an even greater scale...that chance we had is gone.

Besides, even if we somehow stopped all fishing without managing to starve ourselves, and eliminated plastic and chemical pollution, marine life in our oceans would still be struggling to deal with climate change. This is because of the increased acidity of the water, the oceans absorbing so much carbon dioxide from our emissions, as well as due to generally warming water temperatures.

This warmer, more saturated water can hold less and less oxygen, which is a big problem for the animals that live there. As the oxygen levels go down, current "dead zones" will expand, and new ones will begin forming. These dead zones are areas of the ocean where the water quality is just too bad to sustain life. Further pollution will only add to that devastation and accelerate the process.

People and politicians have been talking a good game about saving our oceans for a long time now. But, just like saving the atmosphere - or anything else really - very little has actually been done. Some smaller nations have reduced fishing, but really, unless the populations of Asia were to take meaningful action, everything else is just for show. If anything, both fishing and pollution have steadily increased, and sooner than you might think, that will mean no more fish at all.

Deforestation

From the vast boreal forests in Siberia and Canada, the temperate woodlands of North America and Europe, and all tropical jungles and rainforests that wrap around the Earth's equator, forests occupy nearly a third of the land surface on our planet. True bastions of biodiversity, they help both people and animals thrive and survive. Forests provide a habitat for birds, mammals, reptiles, and insects of all kinds, all of which rely on the forest for their very lives. More than three-quarters of the world's life on land exists within forests. Forested areas also play a critical role in mitigating climate change because they act as a carbon sink, soaking up carbon dioxide that would otherwise be free in the atmosphere to add to the ongoing climate crisis.

But all around the world, forests are disappearing at an incredible rate. These areas are under extreme threat, jeopardizing all the benefits listed above and more. The threats are mostly manifested in the forms of deliberate deforestation and forest degradation. The primary cause of human driven deforestation is industrial agriculture, but poorly planned infrastructure is becoming a big threat too. Forest degradation is primarily caused by bad logging practices or even straight up illegal logging. This deforestation is a double-whammy for biodiversity, as we lose the plants and also the creatures they provide a home for.

Around the world, about 35 acres of forest is destroyed every single minute, adding up to about 18.7 million acres of forest annually. That's an area bigger than all of Ireland. Think about that for a minute. Such mass deforestation due to development and agriculture has permanently removed millions of acres of animal habitat and has put many species at risk for complete extinction.

The large-scale destruction of forests isn't exactly a new problem. As humanity began spreading across the globe, they were cutting down trees at an alarming rate, but the problem has gotten much worse in the last few decades. This has been mostly due to the industrial scale clearing of land for agriculture and human development, to feed and house our ever-growing populations. It is one thing to see a group of people logging timber with saws and axes, but quite another to witness the sheer destructive power of our modern mechanized land clearing machinery laying waste to large swaths of forested land.

Developing countries, or the so-called "third world," have specifically seen most of the deforestation activity. Governments in such nations generally have little environmental regulation and oversight of industries, nor do they have the desire or political will to see such productive enterprises stopped. Take something like slash and burn farming, which is a method by which existing plants and trees are cut down and then burned off to rapidly clear land. This alone has cleared vast areas of tropical rainforest, such as in the Amazon, destroying many habitats rich with biodiversity. Additionally, rainforest soils tend to be quite poor and acidic, which means that the newly cleared land is only good for a few seasons before all the nutrients are extracted and even more forest must be cut down. Cutting down large parts of the Amazon rainforest for cattle ranching has contributed greatly to both biodiversity loss and the acceleration of climate change. In every nation that is developing you will find a shared contribution to the massive loss of the Earth's forests. Such countries do not have the level of national resources to pick and choose what to preserve and what to exploit. They use whatever they have, however they can.

That's not to say the problem doesn't exist in more industrialized nations. Logging, whether with a permit or illegally, is still responsible for millions of acres of forest being destroyed each year in the nations of North America. And several years of record forest fires have burned down millions of acres along the west coast of the United States and much of Australia, in addition to other places like Siberia. This is a trend that will only get more prominent as global temperatures continue to rise and droughts grow more severe.

Suburban sprawl, which destroyed a lot of natural habitats across the world since the 20th century, has begun pushing people out into the last few areas of wilderness in between cities. All of those roads and highways, homes and shopping centers, that come with human development has not only devastated the amount and quality of forested land but has also partitioned it, breaking up larger forested areas into smaller fragments of woodlands. As the human population continues to grow and expand, demand for food and space increases as well, putting ever more increasing pressure on the Earth's few remaining forests.

Deliberate deforestation on such a massive scale is obviously a disaster for our ecosystems in many ways. The climate crisis will also be accelerated with the current rate of deforestation. Forests act as "carbon sinks," absorbing carbon dioxide that would otherwise be in the atmosphere, similar to how our oceans do. As more and more forested areas are laid to waste, the carbon dioxide in the atmosphere will have no other place to go and will instead remain and exacerbate the catastrophic problem of global warming. If forests are replaced with large-scale cattle ranching, like in the Amazon, that can vastly increase the warming effect on the atmosphere. Cattle ranching gives off quite a bit of pollution, both as water runoff and also methane gas. Methane is a greenhouse gas that holds about 30 times more heat in the atmosphere than even the dreaded carbon dioxide.

Urban expansion at the expense of wooded areas will also start feeling the heat more and more. Trees retain a lot of water, which helps keep the air around forested areas cool. When those trees are replaced by the steel and concrete jungle of buildings and highways, cities become "heat islands" which retain and generate massive amounts of heat, causing them to grow dangerously hot in the summer. And it continues to get worse, as hotter air is better able to propagate more smog and particulate matter, making air quality dangerous many people, especially the elderly or those with respiratory issues. With the urban sprawls continuing to grow and the climate crisis making summers hotter around the world, the problem will only get worse.

But biggest threat of deforestation is that it is greatly impacting the Earth's biodiversity, causing the extinction of thousands of species. Since trees provide food sources and homes for so many creatures, razing a forest causes massive ripples through the local ecosystem, especially places of high biodiversity, such as the Amazon. Even the loss of the plants alone, over half of all species of trees in the Amazon are at risk for extinction. Flying insects have declined in population by 75%. Something like 40% of all amphibians, 25% of all mammals, and a whopping 60% of primates are now endangered. And it is all due to human activity and humankinds spread across the planet. As more and more of our forests are cleared, the habitats of other creatures vanish as well, leading to losses all across our vast web of interconnected ecosystems. Mass extinctions have happened several times before in Earth's history, but certainly not at the rate that we are currently seeing. Animals are going extinct at a pace of up to 10,000 times faster than before humans roamed the Earth. These things normally take over hundreds of thousands of years, but now we have to come face to face with the grim reality that humans are cutting down forests and destroying ecosystems for ourselves at the expense of every other living thing on the planet.

The Pharmaceutical Hit

Nature has always been a source of vital medications for centuries. But in recent history species die-off caused by human activity is putting this at risk in many ways. From over-harvesting and pollution to deforestation and climate change, biodiversity loss has resulted in many species of critical plants and animals becoming threatened. Plants are an abundant source of both current and new medicines, often providing us with chemical templates for the design of even synthetic drugs. Yet scientists across the globe say unsustainable harvesting of wild medicinal plants is contributing to biodiversity loss and could limit opportunities to source medicines from nature, both now and in the future.

Plants and fungi have provided, or inspired, key pharmaceuticals for global health challenges, including cancer, heart disease, and malaria, and are still valued today in their natural forms as medicines all around the world. Global demand for medicinal plants has threatened many species, contributing greatly to biodiversity loss and depletion of natural resources that are vital for the continued health of humanity.

Today we are losing undiscovered species before we even become aware of them. Pollution, deforestation, over-exploitation of natural resources, spreading invasive species, and land degradation due to urbanization and industrial scale agriculture is the primary cause of most biodiversity loss. And it is all due to unchecked human activity.

And when it comes to insects, the most diverse group of all living creatures, we've scarcely scraped the surface of the potential pharmaceutical use of what is out there. We may have discovered and cataloged over a million species of insects, but there are millions more of them out there. Some that we know of only have a name, we have not even gotten started learning about their ecology, where they live, or even how they interact with other species. Very little of that is fully known. And with every bit of natural habitat that is destroyed, we are certain to be losing many species that are completely unique in the universe.

Insects are found in every habitat on Earth, and in incredibly diverse forms. Many insects have evolved a huge array of biochemical cocktails, which they use to prey on other creatures, or defend themselves from being preyed upon in turn. These include everything from antimicrobial compounds to venom, and there is great potential being found in these substances for antiviral drugs, new antibiotics, and even drugs to fight cancer cells.

So many of the ailments of humanity may have their cures locked into the DNA of these creatures, and yet we are systematically destroying them at an unprecedented rate, rendering many extinct forever. Who knows what we have lost already, and how much more we stand to lose? At a time when new viruses are emerging, more spillover infections are occurring, and more bacteria are developing immunity to our current weapons against them, we are also on the path of wiping out the very sources of cures for current and future generations.

Welcome to the Anthropocene

There can be no doubt, human activity has certainly impacted the world around us, and in a very negative way. Our effect on this world has been so different than anything the Earth has experienced before, and so scientists have given us a new name for our current time period: the Anthropocene. More than just a name, the Anthropocene is a time of dire consequences and stark changes in reality for the species and ecosystems of Earth. Between urban expansion, resource depletion, and climate change, we have started the sixth mass extinction all by ourselves. Who needs a meteor when you have humans?

Even using the most conservative estimates, the extinction rates we are experiencing now are 100 times higher than what would be considered natural. For example, over 400 vertebrate species have gone extinct in the last 100 years. But if you look at the normal "background rate" of extinction, scientists show that such a number of extinctions should have taken over 10,000 years, or more.

So, we have the biggest driver of biodiversity loss on land being habitat destruction and degradation, mostly due to industrialized farming. As for out at sea, the biggest problem lies with rampant overfishing. Climate change in general will play an increasing role as its effects intensify and accelerate over the coming years. Those twin crises of climate change and biodiversity loss are inextricably linked together, as indeed almost all of our current messes are. But in the end, there is only one common denominator: Humanity.

These occurrences, happening all across the world, are just multiple facets of the same core issue. It all boils down to the increasingly relentless pressure we are putting on Earth's biodiversity and all the things in nature that ensure our wellbeing. It is the way we humans are changing not just the Earth's climate, but it's very blueprint for success. We have been tinkering with a very complex system that is beyond our understanding, taking advantage of what it provides and exploiting that mercilessly for our own growth and profit above all else. Like driving a vehicle hard without doing any maintenance, it won't last nearly as long as it should.

The rich diversity of nature provides us with everything we need to live, from the food we eat and the water we drink to the very air we breathe. But it is not just those physical necessities of life alone. There is also the spiritual and emotional health our world provides to us. Those many individual moments of personal stress relief spent exploring the forests and mountains, swimming in the oceans and rivers, or even just relaxing to the simple sounds of crickets chirping on a quiet evening.

All of us have always just assumed and believed that nature would be here for us and our children forever. We take it for granted that, no matter what else might befall us in our lives, there would always be that embrace by the natural world we could go out and experience to sooth our stresses and bring calm to our chaotic lives. But now we are starting to doubt that. More and more we are becoming aware of how our boundless consumption, shortsighted reliance on cheap fossil fuel energy, and recklessly unsustainable use of natural resources now seriously threatens our future.

Environmental activists, scientific researchers, and even indigenous peoples around the world have been ringing the alarm for decades. But we were not really listening. No one wanted to hear it, we had lives to live, goals to achieve, and dreams to bring to fruition. Very few had time to worry about the problems that seemed so far in the future. The alarms were brushed off, and many of us just assumed that science and technology would find a solution by the time things got to a critical point.

But guess what? We are at that point now. Our understanding of the overexploitation and abuse of the planet has advanced with grim, sharp clarity over that time, and no solutions are on the near horizon. The time is coming due to pay for all those extravagances we enjoyed for so long and pay for them we certainly will.

As far as biodiversity goes, we have now entered an era of rapidly accelerating species extinction and are facing the swift and irreversible loss of many plant and animal species, and with them their habitats and the vital crops we need to survive. Even worse, we are at the same time coming face to face with the disastrous impacts of climate change and other serious problems.

The Anthropocene extinction is, without a doubt, being caused by humans, and climate change, deforestation, habitat destruction, toxic pollution, industrial agriculture, urban expansion, and overfishing are all playing a hand. We are actively killing everything around us, and while that may sound over the top, there is really no better way to put it. And no matter how loud the warning cries, just like with all of our other unfolding disasters, the scientific alarms on biodiversity loss have gone largely unheeded while the problem intensifies. We are indeed killing everything else that lives on this planet, and in doing so we are also killing ourselves.

Hothouse Earth

The recent IPCC report we have already gone over lays out some very simple facts. One key takeaway is that keeping global warming down to within 1.5° to 2°C, which is already bad in and of itself, will actually be much more difficult than previously thought. And you have to also keep in mind that these assessments are made with the expectation that world governments are actually going to do all the things recommended to them. Which they actually won't. In fact, they never have, not once. So, saying that the things will be 'much more difficult' to fix is a gross understatement. Nothing was done when it would have been much easier. Why think they will do anything now that it is difficult?

Another international team of scientists did some of their own studies and published a paper in the Proceedings of the National Academy of Sciences (PNAS). It was titled "Trajectories of the Earth System in the Anthropocene." Herein the authors introduce some new ideas, and their work shows that even if the carbon emission reductions called for in the now doomed Paris Agreement are met, there is still quite a significant risk of the Earth transitioning into a state that they call "Hothouse Earth" condition.

This new "Hothouse Earth" climate will end up stabilizing at a global average of 4-5°C higher than the pre-industrial average, and with sea level up 10-60 meters higher than it is today.

Given what we already know about the effects of the 1.5°C rise, imagine a 4 or 5°C rise. It's not a pretty picture, and while I won't paint the entire thing here, I will touch on that in a moment. First, you should consider that this scenario represents a "stabilized" pathway into the future. One in which we have done everything we can to mitigate the effects of climate change.

Our direct human-caused emissions of greenhouse gas are not the only thing that contributes to the rising temperature on Earth. That above referenced study, and several others, show that human-induced global warming of 1.5 - 2°C will probably trigger what are called "feedback loops" in various Earth system processes, and those feedbacks will drive even more rapid warming. And that is even if we stop emitting greenhouse gases now.

The authors of the Hothouse Earth study identify and examine ten of these natural feedback loops, and therein detail several "tipping points" that lead to abrupt and catastrophic change once a certain threshold has been crossed. Each of these items on the checklist could go from being beneficial to our world as carbon sinks, to becoming drivers of climate change acceleration by emitting even more greenhouse gasses. And such emission would by then be uncontrollable by humans, once the tipping points are passed.

The items in question are the thawing of the world's permafrost, Amazon rainforest dieback, loss of methane hydrates from the sea floor, weakening land and ocean carbon sinks, increasing bacterial respiration in the oceans, reduction of northern hemisphere snow cover, boreal forest dieback, loss of Arctic Sea ice, and melting of Antarctic Sea ice and polar ice sheets.

A good example is Amazon rainforest dieback. The great Amazon rainforest, once a region known as "the lungs of the world" but horribly disfigured by decades of deforestation. Apparently, we treated our lungs the way a smoker does, and now that forest emits more carbon than it absorbs. Not just from wildfires either, but all the time. That tipping point has tipped, and there is no going back now. The passing of the threshold for that tipping point was just recently proven, and yet we are still slashing and burning our way through it.

That's human civilization in a nutshell right there. Smoke 'em if you got 'em.

These various tipping points in the Earth's systems can potentially cause a cascade of failure in those systems. As each one is tipped past the threshold, the effects are accelerated that push Earth towards the next. It's like a Jenga tower, and each block we remove makes it that much more unstable, until one block too many is taken out, or the right one nudged in the right place, and suddenly the whole thing comes toppling down. It is almost impossible to stop the process once it has begun, like a runaway nuclear reaction. Many places on in our world will simply become uninhabitable if "Hothouse Earth" comes to pass.

Back to our other tipping points. These ten distinct facets of what have been called the "Earth System" all have the potential to switch from neutral or helpful to harmful, and with that switch begin dumping more greenhouse gasses like carbon dioxide and methane into the atmosphere than all human activity combined has done.

Most of these Facets of our planetary system have their own temperature tipping points which, once passed, result in the uncontrolled release of these planet-warming gases that would be irreversible. The feedback process becomes self-perpetuating after that critical threshold is crossed for each one.

Carbon sinks are another facet of the weakening system. Earth's forests and oceans have, over the last several decades, absorbed more than half of carbon pollution created by humans, and that is even while those emissions continued to grow. But our forests are shrinking rapidly. Deforestation is a major problem. We keep clearing land for more agricultural use and more urban sprawl construction, not to mention the historic levels of wildfires the world has seen in the last few years. Our oceans as well are showing signs of becoming saturated with absorbed carbon, according to recent studies, with rising acidity, and ever more pollution. These carbon sinks are weakening and may soon stop working altogether.

The melting of the world's permafrost is a big one. Turns out, it may not be so "perma" after all. Methane and CO2 trapped in the long-frozen grounds of Russia, Canada and northern Europe is roughly equivalent to about 15 years' worth of greenhouse gas emissions at today's levels. Were it to really start melting, the release of these gases would dramatically speed up global warming. Similarly, there are rock-like formations in ocean waters called methane hydrates, which were suspected culprits of rapid global warming episodes millions of years ago. These are also vulnerable to global warming, but at what threshold remains completely unknown.

Dramatically shrinking polar sea ice, especially in the Arctic, and the melting of Antarctica's and Greenland's icesheets, means the deep blue ocean water that takes its place absorbs much more heat that was previously reflected back into space by all that icy whiteness. The dreaded "Blue Ocean Event," which could come to pass any year now from the sea ice, and the incredible sea level rise from the demise of the ice sheets. Experts disagree on how much warming it will take to begin melting the West Antarctic and Greenland ice sheets and how quickly they would decay, but all agree that such a tipping point exists, with estimates ranging from 1 C to 3 C. We are over 1 C now...

Remember the 10 to 60-meter sea level rise we talked about? The consequences of such a thing for humanity and civilization would be cataclysmic. Over 60% of the world's biggest cities are less than 10 meters above sea level, as is much of the agricultural land that keeps them fed. All that goes underwater at the low end of our sea level rise spectrum. Just West Antarctica's and Greenland's ice alone would lift the ocean surface by 13 meters. Another 12 meters of sea level rise is locked in the East Antarctic Ice Sheet, which is far more susceptible to climate change than we previously thought. So, it doesn't all have to melt, half does the trick just fine.

There is also the cascading effect of it all. These processes are all interconnected, and the collapse of one will trigger others, which just accelerates faster and faster. The risk of tipping points cascading across the board could come near a 2 C temperature rise, or even a bit before, and will increase sharply beyond that point. This rapid chain reaction of events will change the Earth into an entirely new and hostile living environment.

And let us not forget how all of these climate effects, as they are beginning to press upon humanity and civilization, will greatly stress our other systems as well. Nothing happens in a vacuum. Resource depletion, overfishing of the oceans, political tensions between nations, civil unrest, economic crash…it's a long list. But even just on its own merits, one must consider the fact that the carrying capacity of a 4°C or 5°C degree world, climate considerations alone, could drop down to a billion people or less. That means seven billion of the people here now…are gone.

Finally, as one of the scarier things, while we know for a fact the thresholds exist, we don't really know exactly where the lines are exactly. We can't countdown to the exact moment of the tipping over points. Which means, while it won't be a strategic surprise, since we do know it's coming, it could still be something of a tactical one. But most likely the process begins somewhere in that 2°C range. And also, we have to keep in mind that most scientific estimates are just best guesses based on what we know now, which is far from everything. Look how much more we know today than we did just 10 years ago. We really can't say with any degree of certainty exactly how fast the cascade could be once it begins…and the Amazon has already tipped.

<u>My View of the Approaching Climate Crisis.</u>

I tend to try and simplify things down a bit, and the way I see it is less complex than the more intellectual people out there. To me, the climate crisis is like a speeding truck that we see in our rear-view mirror at the top of a hill a little way off behind us. Sure, it is speeding dangerously, but the distance makes it seem like an easily avoidable danger. I mean look how far away it is, right? And so, we are not too worried about it. We are quite confident that we will be out of harm's way by the time it reaches us, or we can wait to react until it gets a little closer, and so we continue to advance at our leisurely pace while we turn our attention to the immediate issues that are pressing on us.

We kinda forget about that truck a little bit. The media we consume helps with this, spreading hopium. So far away… But little do we realize that it's beginning to pick up speed as it comes down the hill. We paid no attention to the warnings that "objects in the mirror are closer than they appear," and just kept looking at what is right in front of us, and only now are we suddenly looking back in horror at the looming grill of that out-of-control truck about to smash us apart. But now, after waiting and doing nothing for so long, it may already be too late to swerve…

Right now, humanity has progressed long past denial and entered into a pattern of pure delusion. Take the IPCC report that we have discussed. This is supposed to be the penultimate work that catalogues the science behind climate change and all that will happen as a result of it. And yet, even all of these premier scientists and researchers choose to ignore a very big factor with regards to what can be done.

The final section of the report is the "mitigation" section, and it outlines all of the actions that society can take in order to affect the advancement of climate change in our world. This is where the solutions are supposed to go. But when faced with the task, instead of presenting the factual truth they choose to reconcile the scientific realities with political and economic concerns that are not constrained within the bounds of nature and the laws of physics. They have made their solutions about what is economically feasible and politically possible, attempting to bend scientific fact to fit into a human societal narrative. Even worse, they also assume the future existence of technologies, such as carbon capture, which are currently unavailable and possibly unworkable. What happens if those technological wonders are never achieved?

These scientists have made themselves just as blind as the politicians to the truths of sustainability and have instead embraced that civilization can simply continue to exist as it always has in search of continued growth and expansion. Instead of acknowledging the reality about the physical limitations of the planet's carrying capacity, they bow to the desires of the powers that be and look for ways that we can clean up the mess that we have made, while at the same time continuing to make more messes. They chose to work only within the bounds of political reality and tried to make the science fit within that limited space, and as a result, much of the science got left out. The real solutions of degrowth in civilization and the downscaling of economic activity and resource consumption to sustainable levels, well, those ideas are just crazy, right? No political leader could ever make such choices, they would be run out of office with a quickness. Because the simple truth is that such solutions mean doing a great deal of damage to our current civilization and way of life, and there is no way around that. And if we don't take that hit now, the damage will be exponentially greater down the road.

Sometimes you have to rebreak a bone in order to set it so that it can heal properly. There is no way around that. It can be painful and damaging, but it must be done. The IPCC was supposed to deliver the bad news, and then tell us a real way to keep it from getting worse, like a doctor delivering a cancer diagnosis to a patient that will not survive without radical treatment, and maybe not even then. And then the solutions were not supposed to be delivered in terms of what we can do, if we agree, but what we must do, whether we want to or not. And there they failed. Rather than speak truth to power, they bowed their heads instead and told them what they wanted to hear.

And this is why nothing will ever really be done, not anything significant anyway. Because this attitude is reflected all throughout society and humanity. It is not that we are all denying what is happening, it is that we can't see a way to do the impossible, and so we just give up and take the blue pill. We don't want to hear the unsettling and life-changing truth, and so we just put it out of our minds and go on living in contented ignorance, happy with our goals and dreams of a future which will never come. We cannot do anything about the problems that exist while at the same time striving to continue living in a world that does not reflect scientific reality. To take my earlier metaphor to a higher level, we can't effect change because we don't know the truth that *there is no spoon*.

Climate change is real. And it is so much worse than most people realize. Even those among our best and brightest have blinded themselves to that reality, willfully chosen to work and struggle, to sacrifice and protest for changes that will do little real good and be almost meaningless in the end. And we all follow right along and believe them, because we are supposed to believe them. They are our leaders and our protectors, our teachers and our mentors. They are the great minds and deep thinkers that are supposed to lead us in the darkness and show us the way forward. But they are failing, and they missed the correct path a while ago. Now we are all lost in the dark.

That is the real danger of hopeful solutions right there. We think of all these grand ideas all the time. Take protest, for example. People will say, "Oh, if we could just get the majority of people to protest, to lie flat and refuse to go to work or anything for just a month or two, then the government would have to take us seriously!" Well, that may be true, but the reality is that we cannot get that many people to do it, that's just not going to happen. Same with radical restructuring of our political or economic systems. Does it need to be done? For sure. But can it be done, especially in time to actually mean anything? And without sparking unrest and wars? Not a chance.

There is no time for grand solutions, if they were even realistic in the first place. The time for that was about a half a century ago. So, to continue talking about it now, to waste even more time and resources on the unrealistic hopes of unfeasible solutions is ridiculous. The cancer is here, and it is stage four, non-operable. That is all. There is to it. Hope might bring one a measure of peace and happiness in the time we have left, but it will do nothing to improve one's chances of long-term survival. Trying to save civilization is a non-starter. We need to focus on saving our lives.

4

THE FUTURE OF PANDEMICS

The Second C: Contagion

COVID-19 was quite the powerful demonstration that a pandemic is so much more than just a crisis of health. It did a very good job of illustrating the interconnectedness between biology and every other facet of our society, from economics and politics to social and cultural factors, and even conflict. It revealed the inherent weaknesses in many of our societal systems, from government all the way down to our individual interactions with each other. It also exposed the intimate links between humans, animals, and our planet.

As bad as Covid-19 was, it certainly could have been a lot worse. And the kicker about that statement, is that it will be a lot worse in the long run. Because covid is hardly the last disease that will emerge in the world.

Look at me with my "has been" and "was" and all that past tense stuff. We are not even close to done dealing with COVID yet, although the government and media has done a superb job of moving us down the line so we can get back to the real important work of feeding economic growth.

Researchers and scientists have begun warning us lately about how climate change is fueling what could be an incredibly devastating wave of disease and sickness. That wave will be a danger to all people and probably spark new pandemics in the very near future. There will be many thousands of instances of viruses making the leap between various species in the coming decades, and that could result in many unforeseen reactions, mutations, and recombination in these viruses.

As climate change warms the planet, it will cause many animal species to be forced into new areas to find suitable conditions for their needs. They will bring their parasites and pathogens with them, exposing those diseases between species that have never really interacted before. This will greatly increase the chances of what is called 'zoonotic spillover'. Basically, this is where viruses transfer from animals to humans. And that has the potential to trigger another pandemic of the magnitude of Covid-19. Or one quite a bit worse.

Ecosystems the world over are already being disrupted drastically. And as those ecosystems change, the character of disease will change too. Species interactions are already taking place that have never happened before, and those interactions have likely already started the process of spreading and sharing diseases. The time has never been more ripe for increasing events of novel disease emergence, and the coming decades are only going to make the conditions even better for it.

Even if there were some drastic actions taken to combat these changes due to the climate crisis, that won't do a thing to mitigate the risk of such viral spillover. It is already in motion, and there is no real way to prevent it even in the very best-case scenarios regarding climate change. Like with all the threats we currently face, prevention is not possible, the only thing we can engage in now is preparation.

I have read quite a few studies and research papers over the last couple years, and they basically all come down to the same things. There are already thousands of viruses capable of making the jump to humans out there in wild animal populations. For the most part, instances of infections that cross the line from animals to humans have been uncommon, but not only is climate change driving these animals into closer contact with human populations, but we are also actively moving ourselves there. More and more, people continue to push into parts of the world that encroach into animal habitats. Usually due to the expansion of urbanized areas, or the development of new agricultural lands.

There was one study published recently by Colin J. Carlson, and Gregory F. Albery, which is a very interesting read. Titled "Climate change increases cross-species viral transmission risk," this study forecast the geographical movements of over 3,000 animal species due to climate change and land use factors. The forecast progresses until 2070 and found that even under a relatively low level of global warming there will be at least 15,000 cross-species infection events of one or more viruses during that time span. Even more alarmingly, the paper notes that holding global warming to under 2 degrees Celsius won't do a thing to reduce that number. The process is already underway.

There is no answer for that one. Only the hope that we can begin to build health systems and procedures that can cope with the results, and if you have observed the world response to COVID-19, you may get an idea of how slim a hope that really is.

The risks of future pandemics are largely underestimated, and when you look at our actions to prepare for new outbreaks it is plain that we are grossly unprepared. The spread and severity of COVID-19 was a surprise to everyone, which is a clear case in point. And yet all the research shows that future pandemic risks are quite significant, especially with accelerating climate change and growing population densities.

COVID-19 is widely viewed as being a "once in a lifetime" or "once in a century" pandemic, and the mainstream media likes to use such language as they do with everything, the purpose being to keep people from being too alarmed. But much of scientific modeling and research work, based on both historical data and newly discovered information, shows that this is not truly the case.

The real story is that the next pandemic could come much sooner than we think and more turn out more severe than we ever expected before. The stories in the news tend to be downplayed and kept short but let us remember that COVID-19 is not the only outbreak of this decade; we've actually seen a major increase in epidemiological events. From SARS and Swine Flu to Ebola and all the variations of Bird Flu, the frequency and severity of zoonotic spillover infections directly from wildlife hosts to humans has been steadily increasing over the years.

Recently, many more researchers have begun to study and model the increase in risks. Their results are quite worrying and put those risks much higher than many of them originally expected. In simple terms, the general estimate puts the annual probability of a pandemic on the scale of COVID-19 in any given year to be between 2.5% to 3.3%. That means there is a 47% to 57% chance of another global pandemic that is at least as deadly as COVID in the next 25 years. It is not really a matter of if, but when.

Enter Monkeypox

Already!? We are not even done dealing with COVID-19, we even stopped giving all the variants names because there are too many dropping too fast, and just as we are gearing up for a new wave of that little bugger, something else has hit the scene.

And it hit the scene very strangely, appearing suddenly and very widespread across many countries where it had not been detected before, or at least not very recently. Places in Europe, North America, and even Down Under. It also emerged in larger numbers than usual. How it pulled off that feat, we do not yet know at this time.

The strain of the virus in the current monkeypox outbreak likely diverged from the monkeypox virus that caused a 2018-19 Nigerian outbreak, and it has far more mutations than would be expected given its type and history. Several of those mutations are for the purpose of increasing transmission, according to a recent study published in Nature Medicine. That study was done by Portugal's National Institute of Health in Lisbon, and they were the first institution to genetically sequence the current strain of monkeypox responsible for the over 3,000 cases of monkeypox in regions that had never seen the virus before until that time. Between then and now, as of this writing, there are 20,638 confirmed cases worldwide, and 20,311 of them are in countries that have not historically reported monkeypox. That's right, only 327 cases come from countries with a history of the disease. This information is according to the CDCs Monkeypox Outbreak Global Map as of 27 Jul 2022 5:00 PM EDT. That happened pretty quick. No big deal, right?

When the researchers investigated the spread of this infectious disease, one area they look at is the genetic sequences of the pathogen. But there has been a bit of a hiccup when it comes to the monkeypox virus. It seems to have mutated at a rate far faster than it should have.

DNA viruses, particularly the kind with relatively big genomes like monkeypox, generally accrue mutations much more slowly than an RNA virus like SARS-CoV-2, which caused COVID-19. That means that examining the sequences might be less fruitful in terms of tracking how the virus is spreading from person to person. The usual pattern of there being fewer changes to the virus' genome makes it harder to shine a light on transmission factors.

But as researchers around the world started sharing sequences from the current monkeypox outbreak, there was a surprise waiting: There are way more mutations than expected, more than would have been thought possible. 50 or so, in fact.

So many mutations in such a short amount of time seems pretty worrisome, as it meant the virus was evolving to spread more efficiently among people. Researchers found the current strain diverges from the original West African strain of the virus by 50 different mutations, and several of them made the virus more transmissible. This number is far more (roughly six to twelve times more) than they would expect considering previous estimates of the mutation rate for such poxviruses typically have only 1 to 2 genetic substitutions per year. Based on normal evolutionary timelines, scientists would expect a virus like monkeypox to pick up that many mutations over perhaps 50 years, but certainly not in four, which was the length of time between the last small outbreak of this virus and the new one.

Having a large number of mutations is bad, for a variety of reasons. One idea is that the virus has changed so much because it's grown more fit and gotten better at transmitting among people. Monkeypox, unlike something like SARS-2, has historically not been a particularly efficient person-to-person spreader. We tend to see mutations as the result of random mistakes that occur as genetic material is copied. Some mutations don't have any real effect on the virus, some can even be harmful to it, and some can give it an advantage over other strains. It is evolution, after all.

But changes to viral genomes happen as a result of other mechanisms as well, and there are clues this could be what is happening with this new monkeypox strain. We just don't know, it is still very early in this new crisis, and we kind of have our hands full. One thing we do know is that these aren't just random collections of mutations, they are mutations of a very specific type. And this time the virus is spreading in nations that have never even experienced it before, at an alarming rate, with the U.S. actually having the highest rate of infection. Who knows exactly what that means? Certainly not me. But as the outbreak grows, and it seems to be doing so rather rapidly, more infections will be detected, giving scientists plenty more genomes to study. That could help them refine their current hypotheses or introduce new ones entirely.

I will leave it to you, dear reader, to sort through the current mess of scientific confusion and garbled conspiracy theories at your leisure.

Another Plague Waiting in the Wings

Emerging and re-emerging diseases have monopolized public health media coverage for much of recent history, and it wasn't just covid. There have been infectious outbreaks such as SARS, bird flu, swine flu, Ebola and now even monkeypox that have captured the attention of all of us around the world.

But, as bad as that is, there is more. Lately many leading health experts have been saying that an even greater threat to public health may be on the menu—antibiotic-resistant bacteria.

The World Health Organization has recently declared that antimicrobial resistance (AMR) is one of the top 10 global public health threats facing humanity. The main drivers in the development of drug-resistant pathogens are the misuse and overuse of antimicrobials and antibiotics by both healthcare professionals, animal agriculture, and even regular people.

The rapidly progressing emergence and continued spread of drug-resistant bacteria that have acquired new antimicrobial resistance significantly threatens our ability to treat common infections. Especially alarming is the rapid global spread of multi-resistant bacteria (known as "superbugs") that cause infections that are not treatable with existing antibiotics. Our most effective and widely used antibiotics are becoming increasingly ineffective leading to more difficult to treat infections and death from even common infections.

Think about that for a moment. Imagine, going back to the days before we had antibiotics, before the invention of penicillin, when even a simple wound could become infected and potentially life threatening. That is a very scary thought, one which we don't give much attention to now, and it is one of the few disastrous things that have not gotten much media airtime in the last few years. Granted, we have had plenty of other issues to focus on when it comes to world-ending events, but this one is just waiting to raise its ugly head.

Or maybe it already has. Drug-resistant bacteria are already now killing more people each year than either HIV/AIDS or malaria. These deadly new strains of resistant bacteria are causing untreatable infections, fatal pneumonia, relentless urinary tract infections, gangrenous wounds and fatal cases of sepsis, among many other conditions. Just in 2019, drug-resistant infections directly killed over a million people and played a significant role in 5 million more deaths.

This threat of antibiotic-resistance occurs when bacteria slowly evolve the ability to survive exposure to even our most powerful antibiotics. Though this resistance occurs naturally over time, the misuse and overuse of such antibiotics in both humans and livestock have accelerated the process quite a bit. Antibiotic resistance is being fueled by our use of antibiotics with a frequency and enthusiasm that borders on addiction.

In most places across the world, antibiotics are used without proper oversight, and often without any oversight at all. This has resulted in their widespread overuse and abuse. There are simple incidents over time such as when people mistakenly take antibiotics to fight infections caused by viruses without thinking, or when they start an antibiotic treatment but do not finish taking all the medication. And then there are more major contributors, like the willful use of antibiotics as growth promoters in livestock. The cumulative results of all of these actions stacking up over time are ineffective medicines and persistent, untreatable infections by bacteria that has developed the ability to fight effectively against our best weapons.

Antimicrobial resistance is one of the biggest threats to our global health right now. What if something like tuberculosis becomes highly antibiotic resistant? About 20% of cases are already there. Or the various types of so-called 'flesh-eating bacteria?' AMR is a threat that should be taken very seriously, and yet not only is there not much that can be done, but there is also very little public awareness of it at all.

Even now deaths from drug-resistant tuberculosis account for about one-third of all antimicrobial resistance deaths worldwide. If drug-resistant forms of the disease rise, it would complicate an already dire situation. Annually, about half a million people fall ill with drug-resistant TB globally. Less than 60% of Antibiotic resistant TB patients are successfully treated, mostly due to high the high mortality rate. Outcomes for individuals with extremely resistant TB are even worse, with approximately only one-third of patients surviving.

Thanks to the antibiotics we have created, tuberculosis had been largely banished from developed western society, although it never quite went away from the rest of the world, we just like to ignore that part.

But now it's making a comeback, and it's gotten worse than ever. We are seeing an alarming increase in cases of tuberculosis resistant to our very best antibiotics, emerging most recently in parts of the world such as India, China and Russia.

Antibiotic resistant tuberculosis has been called "Ebola with wings" and that should sound pretty scary. It is very easily transmitted through a cough or a sneeze, your chances of surviving it are only around 50%, and that is with the very best of medical treatment.

But that's not even scratching the surface of the antibiotic resistance problem. In the United States alone, at least two million people acquire a resistant bacterial infection of some sort each year, and more than 20,000 of them end up dying from it. Hospitals are seeing bacteria such as E coli (the one always in the news contaminating something every month) that are resistant to even our last line of defense drugs.

Like COVID-19 bringing epidemiology into the minds of everyone around the globe almost overnight, so too will there be another bug of a different color soon that will make the nightmare something altogether new.

Pathogens Frozen in Time

Just last year, research scientists announced the discovery of 33 new viruses that were found in the frozen ice and snow samples collected from glaciers. Now, just within a few days of this writing, another scientific study has found almost 1,000 species of new bacteria in similar samples. The various researchers involved analyzed ice cores taken from Tibetan glaciers. These ice cores contain layers of ice that accumulate year after year, trapping whatever was in the atmosphere around them at the time each layer froze, including different forms of bacteria and viruses. Most of those microbes were able to survive because they had remained frozen, and they are unlike any microorganisms that have been cataloged to date.

The big question is, how will these bacteria and viruses respond to climate change? As glaciers and areas of permafrost all over the world are thawing and melting at an alarming rate, the released microbes could travel with the meltwater into rivers and streams and reach populated areas, infecting plants, animals and even people. The results of that could be catastrophic. As many of these bacteria and viruses are very old, some even older than 15,000 years, modern animals and people probably lack any shred of immunity to these new/old pathogens.

For several decades now, the scientific community has observed a trend of glaciers melting faster and faster. In addition, the latest climate models indicate that by the middle of the 21st century, many of these glaciers will have lost much of their structure as it is melted away. As a result, not only will huge volumes of water be released from them, but also all the tiny pathogens that have been trapped for hundreds or even thousands of years. The chances of dangerous microbes escaping into our ecosystem and causing new, previously unknown diseases are increasing rapidly.

It's an old and popular fiction trope used in movies, books, and video game storylines. The idea of scientists out in some frozen, remote hellhole of the world discovering a deadly virus, germ, parasite, or whatever. Inevitably, it gets loose, and the heroes race to try and stop it from spreading around the world, killing everyone or turning them into zombies and whatnot. Usually, they fail. But that's just fiction…right?

Well, now scientists fear the effects of climate change and warming could make that fiction into a reality. Maybe not zombies, but still. Melting glaciers and thawing permafrost could expose humans to bacteria and viruses we've never seen before, and the fear is that some could be deadly. We probably lack immunity, and we might not have the right medicines to fight against them. Some of these "Prehistoric" pathogens could have even been responsible for their own little extinction events long ago, we just don't know yet.

There in an incredible level of microbial diversity within glacial ice, as our new researcher found. Combine this with an increase in ice melt due to climate change, and we see a significant increase in the chances that potentially dangerous pathogens will escape and threaten us all. Newly thawed pathogenic microbes could lead to new diseases and pandemics when they are released into the environment. The evidence suggests that some of these newfound microbes could be very dangerous to humans and other animals. The research team in this case had identified 27,000 potential virulence factors, which are molecules that help bacteria colonize potential hosts, within this new catalog of microbes. They have also warned that around 47% of these virulence factors are ones that have never been seen before, and so there is no way of knowing how harmful the pathogens could be. Could be a new common cold. Could be airborne Ebola. We just do not know, and that is scary because there is not much time left to find out before we have to learn it the hard way.

Many of these potentially pathogenic microbes may not survive for long after escaping the glaciers, but they can certainly still cause major problems. All bacteria have the ability to exchange sections of their DNA with other bacteria, like swapping parts. So even if many of these ice-borne bacteria die shortly after thawing out, they can still pass on some of their virulent characteristics to other normal bacteria they encounter in the open world. This potential interaction between glacier microbes and modern ones could be particularly dangerous. Many of the bacteria we have already defeated with modern medicines could gain new life among us, with new abilities and defenses we are not able to deal with.

There is a great deal that is unknown about all of this, but what we do know is that the threat is real. And, that potential for new pathogens to break free of their icy confines and attack humanity? Well, it could already be well underway.

A Future of Plague and Pestilence

Diseases arising seemingly out of nowhere is one of the deepest and darkest fears of many people. It is one thing to contemplate monsters or the horrors of war, at least we can see those coming. But an invisible enemy, one that kills indiscriminately without regard to any aspects of its target? Something that cannot be there one minute, and then suddenly manifest like a ghost to strike us down? That is a real terror. And the fear of such transcends all lines of culture, race, religion, or worldview. That dread is one area in which we, as a species, have equality.

The risk of deadly contagious outbreak upon the world is one that blends across the entire spectrum of collapse like no other. Climate change can result in increased zoonotic crossover of pathogens from animal species, or even from ancient microbes hidden beneath the ice and permafrost. Conflict can be another way new diseases and plagues can emerge, either due to the conditions of battlefields as soldiers and civilians alike interact across the areas in new way, to even the direct use of engineered bioweapons as a new frontier for warfare. Even the complexity of our human societal systems can provide the spark, as more people begin to live closer together in ever more polluted and toxic environments.

It is not really a matter of whether or not the world will face more contagion as we progress towards collapse, but more just when and how. Just like COVID-10 shocked the world in 2020, so too can a new danger arise rapidly and unexpectedly. The odds of it happening go up every day and, invisible or not, this is a threat we will all see happen in our lives going forward.

5

THE CONFLICT OF NATIONS

The Third C: Conflict

Conflict is unavoidable. Most of us have learned that from the past, although I do occasionally come across cries of "B-b-but, it's the 21st century now!" As if that should make some magical difference in basic human nature. War between villages, cities, and nations have been the norm for many tens of thousands of years, and our time is no different. From the first moment a human picked up a rock to bash another human in the head with and ended up with more food or better mating prospects as a result, violence became a part of what it means to be human. And on a nation-state scale, war is the historical rule, not the exception.

However, things are somewhat different now. Our reach is not only more global, but it has also shrunken down to the level of being able to focus in on the individual hearts and minds of everyone on the globe. Such is our connectedness and informational awareness now, that a shot heard round the world is pretty much any shot at all. If I have a bad experience at a Taco Bell in West Wheatfield, Wichita, with a single tweet I can bring that information to the entire world as fast as the speed of light allows.

In addition to this all-encompassing awareness making good the ability to wage our wars right in the palms of each other's hands, there is also the vast increase in humanity's destructive potential. It is no longer just a world with one or two nuclear powers, no, today we have nine, and those are just the ones we are sure about. Some nations, such as Iran, could have them by the time this goes to print. And not only are they widespread, but most of them are also in the hands of the various powers currently vying for control of international order.

Why fight? Well, coming soon to a reality near you will be the much-fabled Resource Wars, during which countries scramble to control the dwindling resources of our rapidly depleting planet. There are the big ones, such as oil and gas, there are the other ones which people rarely think about that are just as critical, such as food and water, and then there are those special things such as rare-earth elements used extensively in all of our modern high-tech manufacturing.

Many factors have contributed to resource scarcity. The wanton use of fossil fuels has brought us past the peak of our supply, and yet our economies demand the cheap energy just to continue to function. Climate change and pollution have brought the agricultural crisis into the headlines lately, as well as the increasingly severe droughts, punishing both our supply of food and water as well as the ability to produce more for our growing populations.

Nations will inevitably clash over control of these and other resources. In fact, many believe that the opening salvo of the Resource Wars have already been fired.

The Resource Wars

What we are seeing right now is the emergence of the threat of global resource wars. There is an increasing global demand for supplies of energy, food, and strategic minerals that has been sparking intense economic competition among nations in the last few decades. As the years have gone by, more and more national leaders are coming to recognize the impending reality of serious resource scarcity, and that is leading to counterproductive conflicts. Who owns the resources in question, who has the right to develop and exploit them, where will they be sent and put to use, and who controls the transport routes from the natural locations to the final consumers are factors that must be secured by any nation that seeks to survive the coming resource scarcity.

Whether the outcomes result from competition or coercion; from market forces or state command, nations are actively in the process of determining how to achieve and secure acceptable levels of prosperity in the coming decades. Having passed the peak of the world's available natural resources such as fossil fuels and minerals, we are now in a "zero sum world" where no nation can obtain the means to progress without taking them from some other nation. That is inherently a world of constant and open conflict. Even when new sources of supply are opened up, nothing can really cover the amount of what the entire world needs, there is still fear that there is not enough to go around. National leaders know this, Tycoons of industry know this, and the common people are beginning to realize it. There is not going to be enough of everything for everyone to survive, and thus conflict emerges as we begin the Darwinian "survival of the fittest" process on a national level.

The wealth that results from resource development and the expansion of industrial production increases national power just as it uplifts economies and once uplifted the standards of people. This can feed international rivalry on issues that go well beyond economics.

Most people usually think of economics as being merely about business, but the distribution of industry, resources, and technology across the globe is the foundation for the international balance of power between nations. In a world of resource scarcity, the economic issues in a nations foreign policy become weapons of conflict themselves.

Controlling the access to resources can be used as political leverage, as we have seen with Russia and China. They both have demonstrated that, Russia quite recently and dramatically. Indeed, China has long engaged in an aggressive campaign to control global energy supply chains and to protect its monopoly in rare earth elements. Given the recently announced "strategic partnership" between Russia and China, it is clear that Beijing is abandoning its policy of "peaceful rise". This is not an unexpected turn of events given the brutal nature of the Current Chinese regime. Every nation is now looking to these more recent actions, paying close attention to the economic and geopolitical tensions underlining the conflict for natural resources as it begins to play out. Every country in the world is realizing that now is the time when they have to seriously look out for their own national interests and what must be protected to ensure that their people can enjoy a level of peace and prosperity in the future.

This is just the beginning of a global struggle to survive for all nations, and all are beginning to look beyond humanitarian concerns outside their borders, and instead examining what the impact to themselves be from globally depleting resources, climate change and expanding world population as well as the accompanying social unrest and upheaval that go hand in hand with it all.

Competition for, and scarcity of, natural resources is a national security threat to all nations, and on a par with global terrorism, cyber warfare, biological warfare, and nuclear proliferation. In addition to the conflicts between nations sure to erupt in the face of resource scarcity, there are still other non-state actors to contend with. Terrorists, militants, religious extremists, and international crime groups are certain to use declines in local food security, for example, to gain legitimacy and undermine government authority in many nations very soon.

The scarcity of vital resources including energy, water, land, food, and rare earth elements in itself already guarantees geopolitical friction between nations. Now add lone wolves, criminals, and extremists who exploit these scenarios into the mix, and you can see the importance of this is clear. It is readily apparent that threats are more interconnected today than they were, let's say, 20 or 30 years ago. Events which at first seem local and irrelevant can become viral sensations overnight and have the potential to set off transnational disruptions across the world. We saw this dynamic play out during the George Floyd riots and protests, which quickly spread across many parts of the world even though it was a distinctly American issue. At some point, people growing frustrated from lack of government protections against dangers will evolve into perceiving their governments as the danger itself, eventually taking up arms and transition into either a new form of government or dangerous gangs of criminals. Now, we can also look to the farmer revolts that began recently in the Netherlands, but quickly spreading to other nations as well.

And it is not just nations and people that are drawn into conflict, opposed on different sides of the issues, but businesses and corporations as well. We see daily the various obstructive businesses and wealthy individuals trying to get in the way of climate or conservation projects in exchange for increasing their companies' net profits. They will use the financial power at their disposal to command politics go in the way they want to that end, or if they must, they will use it to drag down any policy maker who cannot be managed any other way.

No matter how you slice it, this will be an altogether different kind of conflict from what the world has seen in times past. And it will infect every part of human society. If you look at the history of conflicts and wars and everything else, it is always about resources. Who has them, who doesn't have them, and who wants them, those are the factors. We have had many wars in this world. Humanity is very good at starting them, and I think we as a species have not picked up on the lessons of history and we are very, very naive about the motivations of other countries and cultures and why they do certain things. But this time is different because everyone is starting to see that this is the end of the cheap and plentiful. Things are running out, things that just cannot be remade or found somewhere else, and this conflict is going to be for all the marbles that are left. No more is there any worry about what might come after, or future relations after the wars are over, because that is an entirely new post-degrowth, post-collapse world where every facet of human civilization has been forever changed.

Simply put, there are no reasons anymore to be cautious, or to pull one's punches. This is it. The winners may survive, but the losers surely will not. The beginning of this period of all-encompassing resource wars is playing out right now in Eastern Europe. Coming soon to a nation near you.

The Breaking of the World

As of this writing, the most significant thing to talk about with regards to conflict, is obviously the war raging in Europe at the moment, between Ukraine and Russia. At least officially between Ukraine and Russia. Increasingly it has been taking the form of a proxy war that is NATO vs. Russia in all but name. As some have said recently, we may well already be fighting World War Three, we just have not officially declared it as such just yet. Well, the Pope did recently, I suppose...

Russia's invasion of Ukraine, and its continued assault on western hegemony as part of it, could be described as a pivotal moment in modern history. As the balance of power in the world is forced into a shift, this conflict is a turning point comparable in importance to the fall of the Berlin Wall in 1989, and even to the outbreak and resulting territorial reshuffle of WW2. In fact, it actually bears a lot of resemblance to that great war, in both the manner of its beginning and to world events and reactions of the time. To go a bit further, one can see this as the first war from which a great many others will be spawned, as the shocks from it reverberate throughout the nations of the world.

Whether this ominous view of the war turns out to be justified, only time, and future historians, will be able tell. But there's no doubt that in the violent, tumultuous days after February 24th, the established international order has been shaken, stirred, and in some respects, upended in extraordinary, unexpected, and entirely unwelcome ways. And quite purposefully so. That established order just might be the real target here, and with its end comes either the chaos of multipolarity or else the rise of a new order by which nations must abide.

It was not very long before this war, that the nations of Russia and China put out quite an interesting statement about a new cooperation towards the goal of creating a "new era" in the world, a new order dominated not by western hegemony but by themselves. That statement basically declares the departure of Russia and China from the Western economic system and the dismantling of the world security architecture under "international rules-based order". It is also quite interesting that this statement came out only three weeks before Putin actually pulled the trigger to roll into Ukraine, so I highly doubt the coming war was not a big topic of discussion between himself and Xi Jinping. Recently, it has come out that Russia and China are looking to quickly create their own currency to challenge the dominance of the US Dollar in that regard, and it would be made up of a collection of currencies from the BRICS alliance nations. Interesting move which further reinforces the idea that they are trying to move the rest of the world away from the Euro and US Dollar. Something that the U.S. will violently oppose.

The conflict in Ukraine has become the catalyst of a new era emerging in the world, a trigger for radical upheaval. It has created a bombshell on the world stage that could create a new global geopolitical battle, and result in a much-altered future for us all. As the conflict has drug on, we are seeing exactly that. And if Russia does manage to show the world that, yes, one can still go out and defy the imposed rules in order to conquer one's neighbors, well, like letting criminals get away with robbing banks, you will suddenly find yourself with a lot more bank robbers.

In trying to find meaning behind the sudden press of Vladimir Putin's drive for Ukraine, most people are coming up empty in the "why" column. There are, of course, plenty of reasons stated in the propaganda of both sides, some of which sound pretty good, but all of them fail when one looks at the bigger picture.

Because they all assume this is about Ukraine, and only about Ukraine. NATO encroachment, the defense of Russian-speaking people, desires to reclaim ancestral lands, fears of being cut off, all these things sound okay. Sort of. But the reality is that none of those reasons are worth the consequences of what we have been seeing unfold.

Have you ever played Monopoly? What is the least fun position? It is that of the player who, early in the game, realizes that he is already doomed. The dice did not go in his favor one too many times, and he knows that, as the board stands, he has no chance of winning at all. It happens, and the other players still have lots of fun, because they are still really competing with each other, while the loser grows bored and upset.

The operative phrase in that statement is "as the board stands." But what if there was an accident? "Oh no! I spilled my dinner plate across the board!" "The dog just jumped on everything!" "That guy got his chocolate in my peanut butter!"

Now what? Dang looks like we all have to start over. And that losing player, well, now he is back in the game. A very simplified analogy, but relevant.

Because I believe that this entire thing has very little to do with taking over Ukraine, a very large nation which Russia knows it would be very hard pressed to occupy and govern permanently. I have been writing on various forums for quite some time now about this, and as things stands now, much of what I was predicting has begun to become reality.

What I was originally thinking is that this war was just the opening gambit in a much larger operation to destabilize western hegemony, send the global economy into a crisis, and rewrite the global security architecture away from the currently imposed system of rules-based order and closer to the way things were when wars of conquest were more common. When might still made right. This is not just Russia looking to re-establish its own empire and traditional seat as one of the premier world powers. This is a combined effort by a coalition of nations, a group that is fed up with following the rules that the West saw fit to impose upon the world after WW2. This is about bringing the entire world to its knees, and Western hegemony with it, and resetting a balance of power in the world to one in which nations can take a freer hand in managing their own goals of domination, without any bowing to ideals not their own.

And so far, it looks like that is exactly what has been happening. Russia has continued to both advance it's gains in Ukraine, while at the same time dragging it out to inflict maximum damage on the world through food and energy crises. Iran and Venezuela, long time sufferers under international sanctions, have recently released their own similar "Joint Statement" to declare intense cooperation between each other for the avoidance of such sanctions. Israel is also getting themselves up in arms about Iran's continuing nuclear weapons program, making all the appropriate rumblings a nation makes before going to war. China themselves have finally started to hint at letting the other shoe drop with their recent declaration of continued support for Russia despite the ongoing action in Ukraine, and they have really let the world know who's side they are on. Russia themselves, under heavy sanction, has actually seen oil revenues increase greatly, given the market price hikes we have seen, and the fact that other nations like India and China are buying everything they can from them. The Ruble has actually strengthened quite a bit recently, which probably came as a surprise to the West.

All of this and more is happening, and to the point where even the biased western media has been forced to comment on Russian territorial gains, a previously verboten topic while such things could still be hidden. The eastern and southern portions of Ukraine are well and truly smashed and under Russian control, other than a few pockets of resistance. Since switching away from attempting the US style of ground warfare, the Russians have gone back to their bread and butter of slow and steady advance under withering artillery bombardment, and that is something the Russian Army was designed and trained for, not so much maneuver warfare.

Markets all over the world are in freefall. In the US, we are still talking about recession, as if that wasn't already obvious to people, and full-on economic depression seems likely. Some analysts are actually talking about a complete collapse of the US economy. And that is not even considering what the smaller nations of the world are facing.

Weaponizing Food and Energy.

Russia is embarking on an interesting new tactic in its war on the world: The weaponizing of food and energy supplies. The prices of grain, cooking oil, and other food commodities soar following Moscow's invasion of Ukraine, one of the world's largest wheat producers, and let us not forget that Russia itself is one of the world's largest suppliers as well, supplies it is now withholding.

In the Russian-occupied areas of Ukraine, Russia's army is confiscating grain stocks and machinery, again another tradition of war that the modern westernized ideal frowns upon. Spoils of war, and such. In addition, Russian warships in the Black Sea are blockading Ukrainian ships full of wheat and sunflower seeds, although the official statement is that it is Ukrainian mines that are preventing the movement of those ships.

Russia itself is now hoarding its own food exports as a form of warfare, holding back supplies to increase global prices, or trading wheat in exchange for political support. This is not an entirely new tactic, using hunger and grain to wield power, but it is certainly new in the modern era, and being wielded to great effectiveness in our globalized world of just-in-time supply chains. Seems like someone came up with a very devious idea…just in time.

Perhaps the idea came from the historical precedent of the Soviets' crop seizures and the devastating famine of the 1930s, a famine that killed millions of Ukrainians to be sure, but also succeeded in its political goals. This seems to be something similar, but writ large on a global scale. Moscow has begun weaponizing famine and using it against the rest of the world. Evil, to be sure, but if it works…

Already on close to war footing to counter Russia's westward advances in Ukraine, Europe is now finding itself pressed to prevent a famine-driven mass migration crisis coming to the European shores from North Africa. As Europe faces a two-front, geo-economic war of attrition with Russia with regards to both food and energy, the physical battle continues and ever so slowly gets better for Russia.

When Vladimir Putin gave the greenlight to invade Ukraine on Feb. 24, 2022, he also detonated the financial equivalent of a nuclear bomb within the global economic system in the form of inflation and both food and energy shortage. That cannot be anything but by design. Commodity shortfalls resulting from the war's disruption of supply chains, and the simultaneous limiting of exports by other nations in defense, have sent consumer prices soaring worldwide. The most immediate impact has been felt in food production chains, especially in Africa's food-insecure states on Europe's southern borders. Already overstretched supporting a beleaguered Ukraine, the European Union's immediate concern is that hunger-driven migration from Africa could be more than the EU can handle. For Africa's own concern, a much broader and more violent repeat of the "Arab Spring" is a big worry.

Because remember, it is not just Russia here. Russia is the vanguard for sure, the dirty end of the stick for a larger coalition of nations, doing all the actual damage and absorbing all the hits back, just like a good tank it supposed to do. But let us not forget that China also decided to start hoarding grain supplies…many months before the war actually kicked off. It's almost as if they knew in advance or something. And soon after, India, another of the world's largest exporters, also announced that it would henceforth be restricting and/or ceasing all exports of wheat themselves. Stating concern for their own populations needs, it is very interesting how perfectly this aligns with the Kremlin's own plans for weaponizing food supplies.

That whole BRICS alliance might have been something to pay attention to after all. Just recently, in fact, June 22nd, 2022, Russian President Vladimir Putin stated that Moscow was now firmly in the process of redirecting its trade and oil exports to BRICS countries as a result of Western sanctions. The BRICS group comprises Brazil, Russia, India, China and South Africa, and their economies account for more than 40% of the global population and nearly a quarter of the world's gross domestic product. That's a pretty big chunk of the world that would seem to be in opposition to the western goals against Russia. Not only that but many other nations are now expressing an interest in joining BRICS. Nations such as Argentina, Egypt, Indonesia, Kazakhstan, Nigeria, the UAE, Saudi Arabia, Senegal and Thailand, among others. Saudi Arabia would be a big kick in the groin to the U.S.…

China itself has refused to criticize Russia's war in Ukraine or even to refer to it as an invasion in deference to its ally, and at the same time is has condemned the US-led sanctions against Russia and accused the West of provoking Putin. As part of his comments, Xi Jinping said that the West imposing sanctions could act like a "boomerang" and a "double-edged sword," coming back to smack the sanctioning nations even harder, and that the global community would suffer greatly from "politicizing, mechanizing and weaponizing" global economic trends and financial flows. Meaning, basically, that the global economy is a complex machine that should not have wrenches thrown in it. And it was not so hard to know in advance that sanctioning would be the first and hardest move made by the West in response to war, and what one knows in advance one can plan for and influence. Russia and China knew what was coming, and they turned that weapon back upon its user.

To be clear, Russia is not solely responsible for the global food crisis. The state of world food security was already growing stark, from a variety of factors, both economic and related to climate change and supply chain disruptions due to the COVID-19 pandemic. Russia simply identified a ripe opportunity to begin pushing hard on the vulnerable points in an already fragile global food system and that system started collapsing from its own weaknesses and overly complex nature. The timing probably never would have been better than after the pandemic, and Russia seized the opportunity.

Another factor Russia has taken advantage of is the role of fertilizer. The production of food begins in the soil with fertilizer, the foundation of the industrial agriculture we require. Modern fertilizer created the enormous increases in crop yields that have driven the world's population boom by about 6 billion people over the past 100 years, and without it things can collapse to a point of not being able to feed all of them pretty damn quick. Not being able to feed one's population is a distinct fear for the leaders of many third-world nations, as if your people cannot eat, they will eventually eat you.

As Russia happens to bet the world's largest fertilizer exporter, fertilizer represents one of the greatest vulnerabilities for both Europe and Africa which Russia can also take advantage of. Russia's advantageous position is boosted further by its status as the world's second-largest natural gas producer, since gas is a key and critical component of nitrogen-based fertilizers. Nitrogen is delivered by fertilizers as the nitrogen-hydrogen compounds ammonia and urea. The hydrogen comes from natural gas, and its price accounts for 80% of fertilizer's cost. Even before the war in Ukraine, natural gas prices had been rising sharply. As of Fall in 2021, Europe's natural gas prices had increased by 400% since January of that year. The price spike saw the cost of ammonia rise to $1,000 per ton in November 2021 compared to $110 earlier in the year, almost 10 times the cost. Accordingly, the price of all the most common, and thus most critical, fertilizers rose to their highest level in a decade.

In yet another example of timing being too good to be anything but strategic, the launch of the Russia-Ukraine war jacked prices even higher, essentially turning fertilizer shortages into an existential crisis for global food security. But only for the rest of the world, not, of course, for Russia or China. And those two are the world's two largest fertilizer exporters, accounting for a combined 30% of global exports. Both of them had also already imposed export restrictions on fertilizer, quite conveniently. Two weeks into Russia's invasion, Europe's largest global fertilizer producer, Yara, cut ammonia production at its European plants by more than half because of soaring natural gas prices, and other producers in Europe were forced to shut down entirely. Keep in mind that the world has not yet begun to feel the effects of this in crop yields.

In the face of this, Russia has the political advantage of being able to shrug its shoulders and blame the West for all of it. While the US and Europe accuse Russia of doing exactly what is being done, Russia's UN Ambassador Vassily Nebenzia can dismiss those claims as "absolutely false," plausibly denying "that we want to starve everyone to death." Nebenzia goes on to say that the sheer enormity of the sanctions on Russia have disrupted transportation routes, impeded movement of Russian vessels and banned them from entering ports, caused freight and insurance problems, restricted commercial financial transactions and created difficulties with international monetary transactions. "If you do not want to lift your sanctions of choice, then why are you accusing us of causing this food crisis?" he asked. "Why is it that as a result of your irresponsible geopolitical games, the poorest countries and regions must suffer?"

Why indeed. Seems that turning an enemy's strength against them really is an effective tactic.

So too with energy. Ever since Vladimir Putin took up the mantle of leader in Russia, the world had been watching nervously how Russia may act given its position as a dominant energy supplier. And no one was as nervous as Europe.

Europe's first collective "oh, shit" moment came in the dead of winter of 2009. Russia cut gas supplies to that most important and lucrative European market over a price dispute with Ukraine. Hmmm, Ukraine... The Ukrainian nation has always served as the main hub and key transit territory for carrying gas from vast Russian reserves to all of Europe, from the United Kingdom and Spain to Germany, Poland, and Bulgaria. This was the first time that the world began to realize that the energy supply from Russia could be used as a weapon, and a very effective one.

Strangely, despite this rude awakening, almost nothing has been done to safeguard the world's energy security against future threats from what turned out to be a devious and ruthless supplier of energy. Commanding vast oil and gas reserves, Russia continued to reap the benefits of revenues that fueled its one-sided economy and military modernization. Russia also cultivated an image of pro-business modernization in its external business dealings, thus calming fears and attracting tens of billions of dollars to new gas and oil projects from the Arctic to the Far East. Yet the weaponized design behind its energy projects never really disappeared at all, in fact, it expanded. A good example is the rise of the Nord Stream pipeline, designed to carry gas across the Baltic seabed to Russia's primary customer, Germany. Gerhart Schroeder, the former chancellor of Germany and Putin's longtime friend, was conveniently offered the position as chairman of the Russian state-controlled Nord Stream pipeline company immediately after leaving office in Germany. After that, just very recently in May 2022, he found himself nominated to the board of Gazprom, the Russian majority state-owned multinational energy corporation. From German Chancellor to the guy with his hands around the throat of Germany's energy security, and at the direct beck and call of Vladimir Putin. Very interesting. Just how long a game is really being run here?

Russia's invasion of Ukraine sparked an immediate response from the West following that brutal military action. The resulting economic sanctions included the elimination or drastic reduction of oil and gas imports by the United States and European Union members from Russia. This was an obvious, and easily foreseeable action, there can be no doubt that Russia would not see such an attack on their revenue coming. The entire purpose of sanctions after all, would be to destroy Russia's ability to fund its war in Ukraine. But, despite hurting Russia's much-needed revenue stream, a consequence of the sanctions was a sharp increase in energy prices worldwide, underscoring just how much of the world's energy security was in the hands of Vladimir Putin. This is something that the West, in its haste to "do something" did not anticipate, and the western media services have been working staunchly to keep the results out of the common knowledge of the public.

The invasion, and the ensuing sanctions from the West, have now sent oil and gas prices skyrocketing, lifting gasoline and diesel prices in the United States to unprecedented levels. Natural gas prices also have climbed around the world. It was no accident that Russia invaded in February, when it is coldest and European demand for gas for heating buildings is highest. They knew exactly what they were doing and what would happen. And furthermore, as part of a larger plan, it acted like a swift tug on the leash to Europe. Many no doubt held their breath waiting to see if Putin would just cut it all off entirely, and in doing so they caught a glimpse of what that would look like.

Germany alone could fall into a significant recession if supplies of Russian natural gas and oil are cut off, something that has already been done by Putin to other "unfriendly nations." The German economic powerhouse is heavily reliant on Russian gas, which accounted for 55% of Germany's gas imports in 2021 and 40% of its gas imports in the first quarter of 2022. And like all of Europe, it is not simply a matter of getting it somewhere else. Natural gas moves primarily by pipelines, and such infrastructure takes many years to construct. And simply put, where would they get it? Russia dominates the market in the region, and they certainly cannot make a pipeline from North America. Transport by ship is both costly and only accounts for small amounts when compared to transit by pipeline.

Something to look for, especially this coming winter, will be Russia using its domination of European energy to strike a major blow. It is a weapon he has not yet used extensively, but he certainly has it in his toolbox. That little tug on the leash could be turned into a full-on yank. Cut the flow of natural gas, and European nations will be left with millions of people facing a freezing death.

In Germany, the Economy Minister Robert Habeck recently said that Germany is certainly heading for a gas shortage crisis if Russian supplies remain as low as they are now, and many industries would have to be shut down if there is not enough come winter, just to keep people from freezing to death. "Companies would have to stop production, lay off their workers, supply chains would collapse, people would go into debt to pay their heating bills, that people would become poorer," he said. He also confirmed the worst fears that this was obviously part of Putin's plan to divide the nation, and indeed, the world. He also all but confirmed that there really was no solution to this problem.

But just by doing what he has done, Putin has enabled driving the prices of fossil fuel energy sky high across the world. It is a very good move when thinking about the inevitability of sanctions, because now that he cannot deliver as much to the world, he must make it so he can get the highest prices for what he can deliver. Thus, drive up the price, and in doing so actually make even more profits from selling less of that finite resource. Win-win.

A recent study in this month of June 2022 has found that rising oil prices more than offset a decline in Russian export volumes during the first 100 days of the assault on Ukraine. In fact, Russia's oil revenue Has soared despite Western sanctions, or actually because of them, which I believe was part of the design of Russia's plan. Russia's invasion of Ukraine triggered easily expected global condemnation and sanctions aimed at taking a chunk out of Putin's war chest. Yet Russia's earnings from fossil fuels, by far its biggest export, topped what is very likely a record 93 billion euros in revenue from exports of oil, gas and coal. Yes, I said record earnings. The current rate of earning such revenue in only 100 days is unprecedented, and that is because prices are unprecedented, and export volumes are close to their highest levels on record as well. Nations would need these resources, and Putin knows that full well, and there would be no choice but to pay a premium. The only nations balking are the very Western ones that Putin wishes to deal damage to, China and other allies are doing just fine, and Russia is easily funding its war chest better than before.

It is laughable to see nations stop importing Russian caviar and Russian vodka in defiance, because that has only the most negligible effect on their actual revenues. When it comes to Russian fossil fuels, the US can hold back easily, and Europe can try and struggle along, but at the end of the day modern civilization stops working almost immediately without fossil fuels, and Putin knows this. And by doing what he was going to do anyway, invade Ukraine, and then bemoaning the crippling sanctions, he goaded the West into using their best weapon against themselves. And the world has not seen anywhere near the full effect of what that means yet, not even close.

Let me say here that I am not a supporter of Russia or Putin. But like in a murder investigation, the motive is usually determined by figuring out who benefits from everything that happened, because it all happened as part of a plan. It is not by accident. You must be able to look at the situation objectively, and without the blinders of morality or the hatred of evil. Just because something is evil does not mean that it cannot also be effective, and just because a person may be evil that doesn't mean they cannot also be capable of intelligence and rational thought. It just means that ethics and morality do not weigh on the decisions or the actions taken.

That being said, everything that is happening in the world right now as a result of Putin's actions and the invasion of Ukraine must go towards the planned objective. If you limit your thinking to that objective simply being the conquering of Ukraine, you end up in an illogical conundrum. Because in that context, the invasion doesn't really make sense, and that is why so many people are having trouble grasping "why" Putin did what he has been doing.

Look deeper. Look at all the other things that are happening as a result, not just the horrors being inflicted upon the Ukrainian people. The invasion of Ukraine is merely the device being used to accomplish so much more. It is like the pawn forward opener in a much longer and more detailed game. And that game has a variety of other players on other teams. Again, who benefits? China? Certainly, a distracted and depleted NATO goes towards their benefit in their own goals of taking over Taiwan. Certainly, they are gaining unprecedented access to the natural fossil fuel resources of Russia while at the same time pulling those resources away from their Western opposition. Certainly, in seeking to destabilize the West they must have new consumers for their products, and the Russian markets are opening up much wider now...

Look deeper.

A World on Fire

Take a glance through the news out there. More than a glance, do an in-depth analysis. The world is in chaos, people are paying over 5 dollars a gallon for gas in the US, and that's just the average. Many places are seeing 8 and 9 dollars per gallon. Inflation is out of control and markets are crashing, famine is looming across the world, and tensions are rising between nations everywhere. Europe could find themselves cut off from Russian gas supplies at any minute, though probably during winter for maximum effect. It would seem that quite a bit of damage has been done, indeed. And who has to drain themselves propping up everyone? Why, the US, that's who. World Police and whatnot. So, who benefits?

Upsetting the game board, indeed. Which helps put everyone closer to a level playing field again. The sides are being chosen as you read this.

Russia was never going to win the old game of world power and control. Not economically, not politically, and not militarily. It simply wasn't possible. US domination was too great and too entrenched, the game was stale for them. The best they could really hope for was holding their position while others vied for the top spots. So, what does one do? If the goal is to become a dominant player on the global scene, something drastic would have to happen to even make that a possibility in Putin's remaining lifetime. And so, he kicked over the gameboard. The pieces are still falling, not all of it is disrupted yet, but things are toppling fast. And why not? He was losing anyway, so drop it and let the pieces fall where they may.

The interesting thing is that, despite what the western media would have you believe, if you look a little closer you will see that Russia is not all alone and isolated at all. The have other friends in red places.

Ready Player Two

Enter the dragon of China. They are quite a player in the big game of geopolitical power, and economic as well, but catching up to the western leaders of the field is so slow, and so taxing, and Jinping doesn't really have that much of a timeline left to be successful either. They could use a boost. They have their own crises closing in on them, both economically and culturally, and the long game isn't really what it used to be.

China is very well positioned to ride out economic waves, especially if these waves take place mostly within the reserve currency of the US Dollar and those nations that are dependent on it and wrapped up with it. Russia, on the other hand, stands to get hammered quite a bit. But not quite as badly as the rest of the world will in the end. And with a new friend like China...

China makes no bones about wanting to take over Taiwan. They say it every time they get a chance, and recently they even changed the wording of their own laws to allow for, and I quote, "special military operations," to be conducted by the Chinese armed forces. Yes, they said exactly that, which we have heard somewhere before…

Now, these wouldn't be "war" operations, no, just things to protect their own sovereign territory. And the kicker? They have never seen Taiwan as separate from China. Indeed, there is nothing to invade, Taiwan is already theirs, a piece of their sovereign territory, just one currently in rebellion. A rebellion which may need to be put down.

They are going to move on Taiwan, militarily, and soon. Russia is doing the grunt work for the team at the moment, dealing damage, keeping the West occupied and helping to deplete resources. Russia is bearing the brunt of Western resistance and punishment, and also setting the stage to get the rest of the world fired up in smaller conflicts due to food and energy shocks, but when the time is right, maybe even when NATO is fully engaged with the Russian threat, that is when China will drop the hammer on Taiwan.

America is not trying to provoke a new cold war with China, but will act to defend the international order, U.S. Secretary of State Blinken recently said. To quote: "Even as President Putin's war continues, we will remain focused on the most serious long-term challenge to the international order - and that is the one posed by the People's Republic of China."

But the thing is, China is no longer interested in the continuation of that international order, not under the rules dictated by the West. Neither is Russia, or many other non-NATO nations of the world. They want to see the Western view of the world and international relations fundamentally challenged and disrupted. They want the imposed "rules-based" order to be removed from the world stage, so that nations can freely chart their own destinies without outside interference. And Russia and China basically laid all that out for us in the joint statement Vladimir Putin and Xi Jinping made together on February 4th, 2022. Just three very short weeks before the Russian invasion of Ukraine.

The issuance of this statement represents a confident shift in which Russia and China plan to take the lead in international affairs, and it lays out a set of principles and a new, shared view of the world they plan to bring about. Xi Jinping's recent statement that the "relationship even exceeds an alliance in its closeness and effectiveness," seems to indicate that this partnership is solid, and that there is a plan behind it for the long term.

The supremacy ideal of the Russian/Chinese vision is clearly demonstrated by the opening paragraph, Listing "multipolarity" as the first of the "momentous changes" of this coming "new era." The pair of them express their desire for a world order not led by a hegemon that asserts its own standards on a unipolar chess board, meaning the U.S. and its rules. They go on to say that this Western hegemony poses "serious threats to global and regional peace and stability and undermines the stability of the world order."

Heady stuff. The statement also stresses that in this new

era, "a trend has emerged towards redistribution of power in the world" so that each country has a voice that "promotes more democratic international relations." And that is where we get to the most remarkable aspect of the joint statement of all: the emphasis on democracy, and a jab at the American hypocrisy that insists upon its own vision of democracy as correct for other nations but prohibits such democracy between those nations. Basically, America compels other countries to embrace democracy. As an example, the embargo on Cuba cannot be lifted until Cuba becomes a multi-party democracy. But the United States also insists on maintaining a unipolar world in which democracy is denied between nations and America gets to rule them all as an autocrat.

Russia and China have recently begun putting forth the idea of an international order in which all nations have an equal voice. Yet to the contrary, the U.S. has always hypocritically demanded a westernized democracy for other nations while insisting on keeping its unique autocratic role at the international level. The way Russia and China see it, and many other nations as well, America wants to be everyone's boss.

And so, it is at this global level that the two nations have staked out their alternative to that old-world vision. At a time when the war is heating up in Eastern Europe, Russia's own backyard, and tensions escalate in China's region over Taiwan, it really is no small thing when they say their "friendship has no limits."

A Two-Front War

There is a plethora of intractable issues that could spark a hot war between the United States and China, but Taiwan is certainly at the top of the list. And the potential consequences of such a war would be profound, not just for the participants but for the entire world.

For the U.S., allowing China to control Taiwan is a bad ending all around. The issue is not just that Taiwan's tremendous military value poses problems for the U.S. interests in the region. The main issue is that no matter what America does, whether it attempts to keep Taiwan out of Chinese hands or not, it will be forced to run major risks and incur costs in its standoff with China. A Chinese invasion and takeover of Taiwan presents one of the toughest and most dangerous problems in the world. Simply put, America has no good options in that scenario, and a great many bad ones that could result in catastrophe.

A Chinese takeover of Taiwan will shift the military balance of power in the world in any number of ways. Think similar to the Ukraine war. But this time, the scale will be so much larger, the disruption incalculable, and the outcome incredibly uncertain. But, even If China were to take the island unopposed by the U.S. swiftly and easily, that just leaves all of those military assets geared toward a Taiwan campaign now completely freed up to pursue other military objectives. Like the projection of power, similar to what the U.S. has done for so long. And who knows what plans they have for asserting their new dominance of the region. China will also be able to easily assimilate Taiwan's strategic resources, such as whatever military equipment remains, personnel, and the incredibly critical semiconductor industry, all of which would bolster China's global military power. Such would also cripple the economies of the rest of the world and put them at the mercy of China.

My theory is pretty simple. None of that ultra-complicated geopolitical or global economics mess. Almost as simple as a boardgame. Simply, the goal is that the rest of the world will be hurt far more than China and Russia combined by the entire meltdown, a meltdown that they are in fact creating for exactly this purpose. The US and Europe could find themselves knocked down a peg or two by the coming fallout from this war and the sanctions they themselves imposed. Those sanctions can do some damage to the Russian economy, sure, but they have been insulating themselves for years, and the long-term effects are of no concern once the conclusion has been reached that there is no long term. If you and I both lost all our money, well, we are both broke now. But if I had a thousand dollars and you had a million, well, who took the worst loss? And who would be better equipped to handle such a loss, a nation used to every cushy luxury available instantly, or a nation used to the rigors of hardship and loss? Hate to say it, but Russia just doesn't have that far to fall, making the abrupt stop at the end all the less damaging for them.

So many things are going bad for Russia in the short term, but long term? Depends on if you already know that a collapse is going to happen. Perhaps if the world was to continue along the current lines of U.S. dollar dominance and international rules of order, well, then there might be some long-term problems for Russia. But if the entire system is going to collapse? In that case many others have much more to lose.

A coming extreme energy crunch, rampant inflation of the world's number one reserve currency, social and political strife and unrest in places where such is allowed, a worldwide food crisis that could result in millions of famine deaths, economic upheaval in markets across the globe, coming mass migrations due to both famine and climate change...and all of it right after (during?) the worst pandemic of modern history, and also right on the cusp of coming disastrous effects of the climate crisis.

If you were going to throw a wrench into the works of the world, now is the time. And that is exactly what I think the "why" is here. And there are a great many things going on in the world as a result of that wrench being thrown.

Vladimir Putin's ill-disguised threat to go nuclear should the west intervene to halt the invasion has come ominously close to breaking a post-WW2 taboo. And since then, Russian media has been incredibly casual about it. That has undoubtedly inhibited the Western response, with fears expressed about a "third world war". A dangerous precedent has been set. What, really, can the world do to stand up to a bully who also has the ability to start a nuclear holocaust? Saddam had no such ability, and so he dangled from a rope. But one cannot just attack a nuclear superpower. Such is becoming apparent to the rest of the world as well, and this knowledge seems to have sparked a resurgence of interest in nations expanding on, or acquiring, their nuclear arsenals.

Despite some improvement in recent polling, Joe Biden's tenure as US president may be fatally undermined by the war, as well as rampant inflation, rising energy costs, and a slew of other things. He is praised for avoiding direct military confrontation with Russia. But as in Afghanistan last year, he has failed to prevent a humanitarian disaster – or stop Putin. Anger over resulting domestic energy price rises and retail inflation could be his undoing. Or the ridiculous moves by the US Supreme Court to start taking rights away from people again might do it. And the American voters are very fickle things. Biden is all that stands before a red wave that could sweep aside everything come 2024...perhaps even with a little help from someone outside...

China stands to be the big strategic winner if, as seems likely, Ukraine becomes a protracted trial of strength between Russia and the West. They are certainly the winner should the war escalate into a direct confrontation between Russia and NATO, as is starting to look more possible. China's president, Xi Jinping, appears to have given Putin a green light when they met just before the invasion. It is highly unlikely that such a significant event would not have been discussed as, regardless of our "surprise" at its occurrence, this thing had been very long in the making. Now Jinping is backing Russia even more, saying that the partnership has done nothing but grow stronger since the war. Imagine that. China's economy has been hurt by rising commodity costs, but not nearly to the extent that the entirety of the west will be, and that is a small price to pay for increased global dominance.

A 5th Generation for Warfare

Remove your thoughts, just for a moment, from the horrible effects of the war in Ukraine itself. The immediate and televised effects. Let us look at some other little tidbits from around the world. It is there that we will see the cracks appearing in the world of order that we have known for so long.

Every nation in the world, in some form or another, is involved in this war. Or they soon will be. Whether as a victim of shocks to food and energy supplies and prices causing internal disintegration, or as a bearer of the direct economic costs, or even as a future participant gearing up for conflict, the entire world is going to feel this. And there can be no doubt that such was the intent.

Disinformation used as a weapon of war, particularly in the form of "false flag" operations, invented social media "facts" by all sides, and the use of internet bots, has really come of age in the Ukraine conflict. In fact, it would seem that this is the first major conflict to occur in the new age of 5th generation warfare, meaning an information war as well as a physical one. When coupled with cyber warfare, propaganda, media manipulation and rigid censorship, as in Russia, it's a potent means of sowing doubt, division and defeatism. And in general, it has managed to create more varying views of what's actually happening than for any other war in history. The potential for influencing political election processes around the globe is staggering. Russia and China are the champions of this stuff, but America is a very close runner up. To sum it up in terms that are often bandied about with regards to financial markets, nobody really knows "shit about fuck" in this war.

Recep Tayyip Erdogan, Turkey's unpopular authoritarian president and serial invader of Syria and Iraq, is one of several unlikely would-be peacemakers. Erdogan has bought weaponry from Russia, sold drones to Ukraine, and his country belongs to NATO. Is it any wonder why no one trusts him? Turkey has made its own moves regarding Syria, and also spent some political capital to block Sweden and Finland from joining NATO, suspiciously in favor of what Putin wants, and suddenly not very much talked about in the media. High-level talks between Russia and Ukraine were hosted a while ago in Turkey, and while those turned out to be another Russian time-wasting exercise, by hosting them, Erdogan hopes for a boost before difficult elections next year. Somehow, I think he will get it.

Germany's chancellor, Olaf Scholz shocked allies and foes alike shortly after the invasion by suspending the highly prized Nord Stream 2 gas pipeline from Russia and creating a 100bn Euro fund to boost the country's armed forces. That will make them the number 3 highest world defense spenders. For the first time since the Nazi era, Germany has begun to re-arm...and Europe is cheering. Imagine that. But not only is it an unsettling move, it is also a very serious one which would not have been made unless the belief existed that such weapons of war would be needed, and soon. The long period of pacifism in Germany has left them with a very small, very outdated, and very impotent military force. There is only so much money can do to help in the short-term, and I am not so sure there is a long-term…

Famine, and the resulting political and civil unrest, affecting poorer countries in the Middle East, Africa and Asia is a growing fear as Ukraine's and Russia's wheat, grain and vegetable oil exports are cut off. Russia is not allowing Ukraine to export grain out of Odessa, and that makes sense from their point of view, as that grain is soon to become Russian grain once the move to take

Odessa gets underway. This comes at the beginning of a bad run of food crops due to climate change effects, and even worse projections for the future. In Tunisia, symbolic birthplace of the Arab spring revolts, bread prices recently hit an unsustainable 14-year high. In developed nations the pain will be felt as well. And without bread, all we will have are circuses.

Israel is disappointing its friends with its invasion fence-sitting, ostensibly justified by a need to keep on terms with Russia in Syria, but that relationship has been deteriorating fast, especially in the face of an Iran leadership, long supported by Russia, making greater moves of its own in the rush for nuclear weapons. But Israel's rightwing government will be happy if the war scuttles the west's proposed revived nuclear deal with Iran, to which the ever-devious Putin has suddenly raised fresh, and convenient, objections. Recent revelations by Iran itself that it now has the necessary capability to build a nuclear weapon, if it so chooses, has put the pressure on Israel to put up or shut up. Immediately after that announcement, Vladimir Putin made a visit to Iran to solidify ties...and who knows what else. Coordination, perhaps?

Boris Johnson, Britain's now former prime minister, was on the ropes and almost down for the count in the days before the invasion, demonized for his illegal Downing Street partying in breach of Covid lockdown rules. But the war, allowing him to play international man of state, had provided a new lease on political life, at least for a while. He has been one of the loudest critics in Europe of allowing Russia to retain even an inch of Ukrainian territory, certainly doesn't support any peace which allows this, and has been making hawkish statements almost as much as Putin himself. And, not to be insulting, but Churchill he ain't. something he found out recently after being forced to resign as much of his own government began to abandon him. Who comes next? We shall see.

Kaliningrad, the tiny Russian enclave squeezed between Poland and Lithuania, and the three former Soviet Baltic republics are emerging as possible new flashpoints for the coming future. Lithuania recently moved to stop all rail transit across their territory between Russia and its outpost in Kaliningrad, a steep escalation indeed, and a big move which could make the next target of Putin's ire clearer for all to see. Fabricated fears about the well-being of ethnic Russians in Estonia, as another example, have been used in the past to justify Putin's threats, just like in Ukraine. Now they are being whipped up again. Right out of the playbook.

International law and the U.S. imposed "rules-based order" for the world that kept wars to a minimum for so long has taken a beating from which it may not recover. By its actions, Russia has ripped the UN charter to shreds. And the UN security council is powerless to act in the face of Moscow's permanent veto power – which it already used to block a resolution condemning the invasion. Russia also boycotted a hearing on Ukraine at the UN's highest court, the international court of justice in The Hague. The UN has been revealed as toothless, and it brings to mind the workings of the old League of Nations. We all know how that played out.

Emmanuel Macron's oft-mocked vision of a sovereign Europe that maintains strategic autonomy and its own military and security capabilities independent of the US has been given a nice boost by the war. Rattled and fearful EU leaders meeting at the recent Versailles summit agreed Europe urgently needed to be better able to defend itself. But for himself, President Emmanuel Macron's centrist alliance was just denied a straight majority in the National Assembly, while his far-right opponent Le Pen moved up to hold 89 seats. Mo' military, mo' problems. Just what the world needs, right?

NATO has emerged united and stronger, on paper anyway, and while the talk of Finland and Sweden joining up may be on the rocks, it is still floundering around. Cracks in that alliance are beginning to show, however. One should never celebrate a cancer remission too quickly, as early signs of improvement often foreshadow a resurgence of the disease. Currently the US-led alliance is facing criticism for not doing more to help Kyiv. There are also others who believe it is taking things too far. And the war has revived debate over whether NATO's eastward enlargement after the Soviet collapse was a blunder that contributed to the current crisis. Criticism leads to dissatisfaction and that has impact on political elections. Putin may just have to ride things out for a few years and wait for the tides to turn. Who knows what could be in store in the US for 2024, and riding this horse for all the chaos it is worth is all part of the "bigger picture" plan anyway.

Oil and gas are fatal chinks in the western armor when it comes to confronting Russia. The US and Britain decided a while ago to ban all oil imports by year's end. Basically, taking some of the bite out of immediate sanctions teeth. The heavily dependent EU needs more time. Conveniently, so does Russia. But rocketing prices, hitting businesses and consumers, have dramatized how hugely powerful a weapon energy is for Putin. A race to find badly needed "green" and nuclear alternatives has begun. But in the meantime, fossil fuels will be the big winner, as everyone scrambles for more, and climate concerns drop by the wayside. This, of course, means those "peak emissions" dates for the climate crisis will obviously not be met. Should Putin decide to use it, cutting off all gas to Europe would be a potent weapon indeed, especially for winter. Russian winter saved their asses once…

But recently, The EU nations have been quietly backsliding, finding convenient ways to pay for Russian gas in rubles as Putin has demanded. European energy companies appear to have bent to Putin's demand that they purchase natural gas using an elaborate loophole to create a new payment system, a concession that avoids more gas shut offs and also gives Putin a public relations victory while continuing to fund his war effort in Ukraine.

The system, which involves the creation of two accounts at Gazprombank, enables Europe to say it is technically paying for natural gas in euros, while Russia can say it is receiving payment in rubles, which is the requirement Putin imposed on "unfriendly" nations. Because of this, Russia's revenues from the oil sales reached a new record of over 90 billion euros, just in the first 100 days of the war. Who benefits, wasn't that the question to be asked?

Even playing and watching international sports has become a lot harder, especially if you are Russian. The country's athletes and race drivers are among sportspeople banned from European and world competitions. Boycotts have a cultural aspect, too, involving things such as ballet, theatre, orchestras and more. Such unprecedented "virtue signaling" may backfire, by convincing ordinary Russians that they, not just their government, are being targeted. Same can be said of the sanctions, which hit the people directly on an existential level long before they hit the governments responsible. Eventually, one begins to hate the hand that wields the whip rather than the one which invited the punishment.

The quest for truth, which is supposed to be the fundamental purpose of free and independent media, has been further set back by the war. Russia has long persecuted western correspondents. Now it is threatening them with prison if they report openly on the invasion. Facebook and Twitter have been blocked. The EU, in turn, has banned Russian state-backed media channels, deeming them mere propaganda outlets. The concept of the freedom of the press is under siege. And the press, in turn, contributes to its own demise by participating in the spectacle. The Russian media keeps up with their state sanctioned threatening and rhetoric, while the Western media sticks closely as possible to the approved governmental narrative of Ukraine's invincibility in the face of an inept and failing Russian army. And so really, no one is telling anyone the truth. That is something you must find for yourself these days, you cannot depend on getting it from any media.

Record refugee outflows, and an accompanying humanitarian crisis, may overwhelm the ability of EU governments and relief agencies to cope. And this is in advance of the migratory refugee crisis coming as a result of climate change and famine across Africa. Many millions of Ukrainians have fled so far from a population of 44 million. And it is expected to continue growing. Europe opened its borders amid an admirable outpouring of public support. But the EU's longstanding lack of an agreed, collective refugee policy, and Britain's shameful response, suggest troubles ahead as the numbers grow. What, exactly, is going to happen to them? What sort of economic burden will they represent, and how long until that public support becomes simmering resentment?

Sanctions on Russia are the most sweeping and punitive ever imposed. And they were also fired off pretty quickly in a knee-jerk reaction by the West, probably too quickly to really think about the long-term effects, which are even now being turned against them. Not to mention the fact that NATO has pretty much blown its entire non-military arsenal in the opening salvo. What will they threaten with later? Harsh language? Banks, including Russia's central bank, businesses and oligarchs have no doubt been hit hard. The ruble plunged deeply from the get-go. Numerous western brands and companies such as Shell have pulled out. So far Putin has shrugged it off, and that was mostly seen as a bluff. Or it could be the reaction of someone who knows that there is a longer-range plan, and this is something that just has to be weathered for a while. You know the saying, sometimes things have to get worse before they get better. As far as the ruble goes, it is getting better...

And it is not as if the framework behind sanctions is some big secret. Figuring out what the west could do beforehand would be pretty easy, and thus planning to offset it becomes workable. It does not matter how strong your enemy is, if you know what he will do early then you can be proactive in combatting it. If Russia defaults, or retaliates by cutting gas supplies to Europe, the result may be an all-round economic meltdown, big job losses, and a drastic fall in living standards in the UK and elsewhere. The chances of a global economic meltdown grow with every straw we keep placing on that particular camel's back.

Now, will you look at that? Things are getting better. The ruble is actually gaining strength over that which it previously held before the war. China has embraced Russian oil as well, and now Russia has grown to become the leading supplier in the world to China, unseating Saudi Arabia from that position. Just recently, the Russian ruble has continued to rise against the dollar, actually making it the best-performing currency in the world this year. Three months after the ruble's value fell to less than a U.S. penny amid the toughest economic sanctions imposed on a country in history, Russia's currency has mounted a stunning turnaround. The ruble has jumped more than 40% against the US dollar so far.

The primary reason behind the ruble's recovery is the soaring prices of commodities, namely those commodities which Russia has an abundance of. After Russia invaded Ukraine on February 24, already high oil and natural gas prices rose even further. Commodity prices are currently sky-high, and even though there has been a drop in the volume of Russian exports due to all the sanctioning, the increase in commodity prices more than compensates for these drops and does so with Russia expending less resources to boot. Massively increasing exports to India and China help quite a bit as well. Russia is pulling in nearly $30 billion a month from energy exports. Since the end of March, many foreign buyers have complied willingly with the Russian demand to pay for energy in rubles, and that has further pushed up the currency's value.

At the end of the day, our entire global economy is based around fossil fuels, and as usual, he who has the fuels wins.

Taiwan has been watching events in Ukraine with deep unease, of course. Very deep. And initially, shortly after the Russian invasion, the silence from Beijing was deafening. The US refusal to come to Kyiv's aid with direct military support is especially chilling, given the invasion threat the island faces from Beijing. As with Ukraine, Washington has no legal or treaty obligation to fight for Taiwan. Its position is deliberately ambiguous – and inherently unreliable. China is watching the events in Ukraine, too. And Beijing is being anything but silent about things now.

The rhetoric has changed back again recently however, as Defense officials from China and the US are again exchanging heated words and declarations over Taiwan. Washington's concerns about China's "coercive and aggressive actions," in the Taiwan Strait have been the western complaint. These actions include regular military aircraft sorties into Taiwan's Air Defense Identification Zone, and by increasingly larger numbers of aircraft.

For the CCP, Chinese Defense Minister Wei Fenghe reiterated Beijing's determination to "crush" Taiwan's attempt to pursue independence. "If anyone dares to secede Taiwan from China, we will not hesitate to fight. We will fight at all costs, and we definitely will fight to the very end. This is the only choice for China," he said. At all costs and to the very end…seems pretty clear to me what China's position on the issue is, as well as their plans.

Now, we have this new standoff between China and the U.S. as Nancy Pelosi keeps pushing her plan to make an official visit to Taiwan. For no good reason either, just to strut and score points. The very idea demonstrates an ignorance for how eastern thinking and viewpoints differ from those we have in the west. What we may not see as a big deal, China may interpret as the most deadly of offenses. The very idea of such a visit has already pissed off China, which opposes any diplomatic recognition of Taiwan independence. In their view, it is as we would see it should Putin suddenly decide to visit Puerto Rico on a diplomatic mission.

China, which claims the somewhat self-ruled island of Taiwan as part of its sovereign territory, has vehemently opposed this visit in the strongest of terms. Even the Biden administration knows that the move could further antagonize an already mobilizing Beijing, and they have expressed those concerns. But at this point, despite all the risks, it is already too late for Pelosi to back down. If she were to cancel her ill-advised visit, the rest of the world will see that the United States can in fact be threatened successfully. That it can be made to back down in the face of force. In the current efforts by various world powers to prove that exact point to everyone else, the move ties China's hands as well. After all the threats and all the rhetoric spewed, should Pelosi go through with it, China is going to be forced to "put up or shut up." This little play may turn out to be the deciding factor in just when the real kinetic activity is going to begin in this struggle for the defeat of western hegemony. And everyone is watching to see who comes out on top.

Moving on to the drier and hotter nations, the United Arab Emirates is among several western allies in the Middle East and Asia that have failed to show the kind of solidarity with the West that was expected. The UAE has not condemned the Russian invasion, nor has it adopted sanctions against Russia, with which it has close economic ties. Shifty Narendra Modi's "world's largest democracy" of India, is another big disappointment, as is Egypt. These abandonments will not be forgotten and may affect future ties in the west. Additionally, both of them have a neighbor of their own they have issues with, in Pakistan and Ethiopia, and so seeing the lack of opposition to a conquest is a very interesting thing for them both to observe. And does anyone even pay attention to the hell breaking loose in Yemen?

Tensions continue to rise dramatically between Israel and Iran. There have been more of the usual assassinations of Iranian scientists and military personnel by Israel, and Iran has done some seizing of ships and such. More of the same, to be sure, but the talk is growing hotter, faster. Israel has been screaming that Iran is perilously close to developing a nuclear weapon, and Iran has shut off the cameras installed by the United Nations' International Atomic Energy Agency to monitor uranium enrichment. Now, who really knows how close they are or what they are doing, but one thing for sure, Israel is staunchly opposed to Iran gaining nuclear weapons. They have attacked before to stop it, and they will surely do so again. Israel's relations with Russia have grown very strained recently, and Iran has only tightened its ties with that longtime ally.

In fact, just a few days back, the media rumor mill has begun to report on the idea that Iran has agreed to supply hundreds of armed aerial drones to Russia, to help in the campaign against Ukraine. Immediately after, Putin took his first jaunt outside of Russia since the war began and went to Iran. Lots of talks about how nations need to join forces to avoid western sanctions and whatnot. Very interesting. What might the Russians give Iran in return? Well, it's not as if they can get sanctioned any harder if they were, perhaps, to make a little violation of the nuclear nonproliferation treaty…

Venezuela's hard-left government has been on America's naughty list for years. But when US officials visited recently to discuss resumed oil supplies in return for an easing of sanctions, they found a not surprisingly receptive audience. In contrast, when Biden phoned Saudi crown prince, Mohammed bin Salman, himself an avid Putin fan, requesting increased oil production to compensate for banned Russian exports, the prince refused to take the president's call. US-Saudi relations have been circling the drain since Jamal Khashoggi's 2018 murder. This incident will make matters worse.

When it comes to Venezuela, they recently announced a national partnership of their own, with another of the world's most heavily sanctioned nations: Iran. In a recent meeting Iranian and Venezuelan leaders announced their new joint efforts towards "incredible resistance against threats and sanctions by enemies and imperialism." This new "Cooperation Roadmap," the full details of which have been kept under wraps entirely, was signed by the foreign ministers of the two nations in the presence of their respective presidents. Just another pair of the world's largest oil-controlling nations of the world entering into treaty and alliance specifically against western imperialism and sanctions. It would seem that, while the US cannot seem to free itself from fossil fuels, the rest of the world may be ganging up to help them with exactly that…and destroy the US Dollar as a world reserve currency in the process.

All of this plays directly into the idea that this is not really a collection of nations all acting alone, but part of a bigger and more cohesive plan to enable the world to shake off the reliance on the US Dollar like a dog shedding water after a swim, and just as quick and violent. Break the dollar and you break the US economy, and when that breaks so too does the leash the West has managed to keep around the worlds neck since WW2.

In Ukraine, war crimes investigators face an interesting test as evidence mounts of multiple atrocities by Russian forces, exemplified by recent school and hospital bombings. So-called "universal jurisdiction" prosecutions are contemplated in national courts. And the international criminal court has begun investigations. But, just like the US and China, Russia does not recognize the ICC's authority. And if the three biggest superpowers in the world don't recognize it, does it really exist? Or does it just make that court a tool used by them when convenient, but flouted when they are its subject? I certainly haven't seen the US brought up on any charges for so many more bombings in Iraq or Afghanistan. War crimes are only really be punished by the victors upon the defeated, and not until the war is decided anyway.

Besides which, it would seem that much of the western world has somehow forgotten exactly what war really looks like. What we tend to call "war crimes" in today's warm and fuzzy view of how things should be, would historically been looked at by conquerors as some sort of incredible mercy. War is supposed to be bloody, indiscriminate slaughter of the opposing forces that one nation is trying to conquer, including those civilian populations and infrastructure that supports continued opposition. That is the ugly truth of it, such is the historical norm, not the exception. Having any "rules" at all, that is the exception. The West has overcivilized things for quite some time, and has perhaps forgotten this fact, but I do not think the nations in the East ever did. "Scorched earth" is not some desperate act to them, I believe, but more of a viable strategy.

Xinjiang, home to China's persecuted Uyghur Muslim minority, is one of many other global trouble-spots whose urgent problems have been eclipsed by Ukraine, and brushed under the rug by the media, no doubt to Jinping's delight. Millions of Afghans enduring a winter of terror and starvation under Taliban rule also suddenly seem forgotten. The plight of the people caught up in Ethiopia's civil war is another glaring blind spot. And what about the current humanitarian crisis in Somalia, the utter collapse of Sri Lanka, or the doom descending rapidly upon Lebanon and so many other places? Well, in the mainstream media, not a peep, so it would seem that no one really cares…or they don't want you to notice what is really happening in the world outside of your western media bubble.

Younger generations all over the world have good reasons to be confused and to wonder just what the hell is going on. First, they inherited the climate crisis. Then came the pandemic, and the resultant bans on study and travel. Now they face something older generations said would never happen again: a full-scale war in Europe, and probably the entire world very soon. This time, a first, being played out in sordid detail across all social media. They are literally seeing bodies on the ground before the corpses are even cold where they lay. And they watch as both sides of the conflict sensationalize it all. Some social media has recently announced that it is now considered okay to hate people, as long as those people are Russian. There are websites set up so that people can "donate" some money to Ukraine in return for having their message of choice written on the side of an artillery shell or rocket and then fired off into the Russian lines. Seriously, look it up. People are sending birthday wishes and graduation congrats 155mm at a time, to make entertainment out of killing other human beings. 5th generation warfare, indeed. Anyone can now go online, buy a rocket, and watch the video of it being loaded and fired at another group of people.

So many more things could be talked about, and there are so many more boiling little pots of stew to stir, each one on its own close to boiling over with more and more consequences for the world. There are also so many more factors to consider just with the Russia/Ukraine war alone. I don't have the time or the word-count to detail them all here, but hopefully this provides you with a lot of openings for doing your own research. And don't just look at those top Google results. Dig a little deeper. Look at local news sources and stories in other nations of the world. What are they saying in Peru? Argentina? Kuwait?

What is happening in Ukraine is terrible, and unconscionable. But the fact remains, this is not just about Ukraine. Putin may be attacking Ukraine, but he is destroying the world. This is a war on the world itself. The first of its kind. And it is only just beginning. I believe that it is all part of a larger plan to topple the world into a chaotic mess from which those who are prepared can emerge on top. I also believe that it is a plan that has been in the works for a long time and was probably meant to come a little bit later, but the surprise (not going down that rabbit hole yet) onslaught of COVID-19 combined with the resulting global shakeup provided an ideal entry point that could not be ignored, and so timetables were moved up. It may have started in Ukraine, and the Ukrainian people are going to bear an enormous cost, but it most certainly will not end in Ukraine. And the global costs have not even begun to be tallied. Open your eyes and your minds to more than just what is dangled in front of you. Nations are already collapsing. Look to Sri Lanka, Lebanon, and Tunisia. Look at Afghanistan, Bangladesh, and Ecuador. Look everywhere that the media is not showing you, and there you will find the seeds of our own futures.

This is not just a land grab by some aging wannabe dictator. This is Breaking Bad, writ large on a global scale, by many of the world's most powerful assholes with nothing to lose, and the world will not be the same for anyone in the end.

Either way, my theory could be correct, or it could be complete trash. Don't focus on what I believe or do not believe. My beliefs are irrelevant. Why all this conflict is happening doesn't really matter. All that matters is that it is happening, at the worst possible time, and it is escalating all over the world. By design or by accident matters not. What matters is that civilization cannot afford it.

Nuclear Armageddon

The world is right now facing its highest risk of nuclear war ever. The conflict that is currently raging, and the others about to start, are not going to end like many other wars of the past. Now we have nuclear armed aggressors waging war against nations that are either nuclear armed as well or supported by those that are.

Let's look at Russia. If President Putin comes to the conclusion that he has no future, or that the future he envisions for Russia will not come to pass, he may well decide nobody else should have a future either. An existential crisis is exactly that, it means you cease to exist if you lose. And in that case, there has never been a nation in history which has submitted willingly to such a defeat and eventual absorption before first using every weapon at its disposal to avoid such a fate. Do you really think that Saddam Hussein, had he actually had any weapons of mass destruction, would have preferred to dangle from a rope rather than use them in a last desperate attempt to survive? Would Adolph Hitler have died in a grimy bunker if he had possessed the ability to rain hellfire down upon the forces that assailed him? I don't think so, do you? Good or evil, right or wrong, those are not the things that matter in cases such as these. All that matters is that desperate times call for desperate measures, and those who have the option will use them.

This is a new phase for war in the world. Before the dawn of the atomic age, we really didn't have nations with the option of at least taking out their enemy with them in defeat, which is a concept as old as warfare itself. During the height of the Cold War, we came close to having such a thing play out, but for the most part both sides remained at close to parity, and they both still had things to lose. But what happens now, if one side or the other comes to the point where they literally have nothing left to lose?

More people are more aware of the coming collapse than you might think. The wealthy elite are not stupid, they know what they are doing and are aware of the consequences, and they are still grinding hard. Not from ignorance, but from embracing the inevitable and seeking to maintain their power and comfort for the remainder of their lives regardless of collapse. It is the same for national leaders. They all know that this is the "Final Battle," so to speak. Soon, all the old rules and whatnot won't matter anyway. It is no different than Walter White, who had no fear of the potential consequences because he wouldn't live to see them in any case.

Right now, in the world of the powerful and wealthy, this is the "get it while the getting is good" phase. They are literally stripping the corpse of this planet, and they know that whatever they can get now will be all they have in the future, post-collapse. It is the same for nations. Not all of them are going to survive the coming chaos for long, and those that can are gearing up for the fight now. Control of resources will be the key, be it food, water, energy, or raw materials. The leaders, they recognize that the days of slow expansion within the existing rules of international order are over. There is no time left to wait for economic and political change to gradually eke out gains. And that is partly why we are seeing this new movement of nations such as Russia, China, and even Iran to move away from these methods and instead start going for an all-out assault on the world order.

Look at how many "astonishing" things have been done recently. The "unprecedented" actions by nations. Look at BRICS and the new launch of an international reserve currency specifically aimed at upsetting the dominance of the U.S. dollar. Think about the rhetoric and the strong wording of recent statements by China regarding Taiwan. And look at Iran, the first nation to buck the trend and actually deliver military aid to Russia with the gift of armed drones for use in the war with Ukraine. This list goes on and on like all the rest of my lists of crappy things happening. But they all point in the same direction.

War. Maybe even the last real one.

And that is why I really believe we are going to see the use of nuclear weapons soon. Most likely starting with small, tactical low-yield weapons used by Russia against Ukraine, should either the Russian advance falter hard, or else NATO support for the opposition grow too strong. There is not chance of Russia simply giving up. They are playing, as we all are, for all the marbles now. And what happens then? Does NATO step up or back down? The standard Russian doctrine for the use of such weapons is one that views a nuclear war as quite winnable, unlike us here in the west. It is called "Escalate to de-escalate" and it basically means that their view is that it may be the only way to get the opposing side to back off and capitulate, and thus avoid an all-out nuclear exchange. And even then, Russia is vastly more prepared for a strategic nuclear war than we are here in the west. And they have much more to lose in this current conflict, they have no choice but to continue.

There are a great many analysts and soldiers out there much more knowledgeable than I on this subject, and I have spoken with many of them. I encourage you to do some serious research yourself on the subject, far beyond just the tiny bits I have detailed here. The variables in calculating the risks of nuclear war are many and diverse. For now, I will leave it with my own deeply held belief, after many years of my own research and education, that the use of nuclear weapons in war is coming soon.

Take that with as many hefty grains of salt as you feel you

need to. But do the research first for yourself. Take a real read through the Russian military doctrine, especially things like the policy of "escalate to de-escalate." Or maybe consider the works of Dr. Peter Vincent Pry. Dr. Pry is not just another intellectual stuffy shirt. Right now, he is the Executive Director of the Task Force on National and Homeland Security, and also the Director of the United States Nuclear Strategy Forum. He knows a thing or two about how nations like Russia and China view the idea of nuclear war being winnable or not, and at the end of the day it doesn't really matter what the truth actually is, only what the wielder of such weapons *believes it to be*.

The Energy Crisis: A Double Headed Dragon

At the time of this writing the world is currently in the grip of an energy crisis like no other and this time it is a unique one. And make no mistake, all conflicts between nations now are energy conflicts in some form or another. Indeed, there is even the conflict between the energy industries themselves and the governments and peoples of the world. New unprecedented record prices are being set every week across the world for fossil fuels. The costs of gas and diesel have skyrocketed everywhere, natural gas supplies are at a premium, and as energy increases in cost, so too does everything else.

As it stands right, now the world is economically and socially spiraling out of control from massive, rapid increases in all fossil fuel prices. Much of the blame is laid at the feet of conflict and geopolitics, marking the majority of the problem as being fallout from Russia's invasion of Ukraine. But, while that certainly rang the world's bell in terms of energy prices, the war has also laid bare a dilemma: Not only do nations of the world remain incredibly dependent on fossil fuels to provide almost everything needed for civilization to function, but the scramble to make up the shortfall from the war has revealed that we are kind of running out...

What is unique about this energy crunch from all the others is the timing, which makes this problem two-fold.

On the one side, the ugly dragon of demand has risen its head, and the people of the world are calling for both more and cheaper energy. Developing countries continue to develop rapidly, and their people want the lives they were promised so long ago. Populations are still growing, and more people means more power, especially as the technological revolution has made every single item that we own overpowered in some form or another. From Bitcoin mining rigs and constant internet connectivity to the refrigerators in our homes that need Wi-Fi to monitor their contents and let you know when the milk is going bad. The world wants energy more than it ever has, demands it, and it wants it cheap. Politicians and leaders in office who allow prices to surge too high find themselves out of the running next election, or worse yet at the very end of their rope, so to speak.

However, the other side of the equation is a dragon of a different color, the looming beast of global warming. The more we burn in fossil fuels, the stronger that dragon becomes, until eventually it will devour us all. This is a no-win situation. While nations could greatly reduce their vulnerability to the price hikes of the oil and gas markets, and additionally start to fight the acceleration of climate change by shifting to cleaner sources of energy such as wind, solar, and nuclear power, that transition will take decades. We have gone far too long ignoring the playbook for fighting climate change, and we are now beginning to face the consequences of that. We don't have decades anymore.

So, for now, most western governments are just urgently focused on alleviating near-term energy shocks, putting climate issues on the back burner if you will, striving to boost global oil and gas production to replace the sheer mass of it all that they lost to the sanctions placed on Russian exports. Russia has historically exported a very large portion of the supply to western nations, especially in Europe, but now those are being shunned for political reasons by the very nations that are critically reliant on them, especially when it comes to natural gas which is something that cannot easily be gotten from elsewhere. Other hungry, developing nations, as well as the industrial powerhouses of China and India have been happy to take up the slack and absorb quite a bit of the oil that the westerners refused, so burn it will, just not in the west.

However, that is not the full length and breadth of the crisis. The war in Ukraine began almost simultaneously with the release of the IPCC report that was considered a "red alert" for humanity. Our best and brightest were calling out the truth of our dire situation, desperately pleading for change at this the very last of moments. Even as shackled and restricted as the IPCC was, they still did their best to press the importance of the issue. But, in the face of energy scarcity and a time of war, that report was promptly ignored by world leaders. Immediately, they began struggling to shore up supplies precisely at the very moment when scientists were warning the world that we must slash the use of oil, gas and coal to avert irrevocable damage to the planet and the onset of an existential crisis.

That warning is very easy to sum up. In short, all greenhouse gas emissions must peak by no later than 2025 and then they must be reduced by 43% by 2030 at the latest. And this drastic action would not even save us from the effects of a disastrous change in the climate conditions of Earth, oh no. This change is needed just to maybe have a chance to avoid the absolute worst of it.

So, with less than three years left to reach peak emissions, what has been done? We have doubled down on fossil fuels. Yeah. Previously shuttered coal power plants are being brought back online, nuclear plants are actually getting shut down, and world producers of oil are being begged by national leaders to ramp up production in order to try and get a hold of skyrocketing gas and diesel costs.

The reality is that countries have become so consumed by the immediate energy crisis that they are neglecting, or throwing out, longer-term policies to cut reliance on fossil fuels. This is not just a shortsightedness. This is a deliberate decision to put the economic and political needs of today ahead of the survival needs of tomorrow. A decision that will not only set the world up for more oil and gas shocks in the near future, but also usher in the reality of a dangerously overheated planet.

Can Civilization Continue Without Fossil Fuels?

Every conflict is, in a way, an energy conflict. This one is no different. The past three centuries of human progress on this planet have been powered by fossil fuels: coal, oil and gas. Burning too much of what is left of it will lead to an existential catastrophe for humanity. But, at the same time, cutting ourselves free of fossil fuel energy in the short amount of time that remains for us to do something is an economic and political non-starter. There really is no way to save the earth in the long term without giving up on growth and giving up that model of growth would leave us in the midst of societal collapse in the short term.

Catastrophic collapse due to climate change has become a reality no matter how you slice it. We are simply talking about "when" and "how" at this point, not "if."

Many people will continue to fight to reduce and eliminate our use of fossil fuels, but the problem is simply too complex to even comprehend fully. Everything about modern society is based on fossil fuels. It's not just about the economics or geopolitics of it. Literally, every single aspect of our modern way of life depends on the continued use of fossil fuels. Our entire system of functioning rests with fossil fuels, the economic progress we have made and the system of business and industry it supports. The whole global communication system which allows us to visit each other in a completely different way than we did before is based on an economy with access to very cheap, abundant, and efficient energy sources, which have no real viable replacement.

It would have global consequences for transport worldwide, as phasing out fossil fuels rapidly would create a doomsday scenario. And doing it slowly, which is highly viable, is a solution for decades ago. It is simply too late now to take a slow measured approach, and any remotely possible fast measures would result in absolute chaos and societal collapse on its own. There are less than 3 years left before emissions must peak in 2025, and less than 8 before they must be reduced by half by 2030. We have not been able to accomplish so much as a 10% reduction over a period of years before, what chance is there really that we could suddenly manage it now, and in the face of both the civil unrest and world conflicts that would no doubt result?

Fossil fuels are about so much more than just energy, although they still supply about 80% of world power usage. The global economy is centered around them in every way, and almost every product or bit of food you need to live must be produced, packaged, transported and delivered by many forms of fossil fuels. From actually moving goods around the world and fertilizing crops to manufacturing the plastic that is in everything and even pharmaceuticals.

Apparently, most people do not seem to know that pretty much all conventional plastic is made from petroleum products, primarily oil. Additionally, just about every other product is made with some plastic content. Even many pharmaceuticals rely on petroleum for production. I would have thought that this was common knowledge, but I guess not. Seems this information falls into the "useless trivia" category in most people's brains.

I will spare you the full listing of the thousands of daily necessities that are made from petroleum but suffice it to say it is needed for everything from antihistamines to upholstery. Google it for yourself, ask exactly what is made from a barrel of crude oil. The answer might surprise you. You are wearing items right now made from it. You have probably consumed it into your body already today. If you look around you, almost every item you see would not exist without petroleum refinement.

And that is why it cannot just be stopped. Our modern civilization is based entirely upon fossil fuels, in almost every way possible, and such civilization cannot continue without them. There will be no end to fossil fuels in the short term, but they will bring about an end to themselves one way or the other.

What Can Be Done?

There are a lot of pie-in-the-sky ideas out there and setting aside for the moment that most of them are not even technologically or economically possible now, they pretty much all have one other huge flaw. From renewable energy to carbon capture and geoengineering, all of them are trying to marry two completely contradictory objectives. They imagine a future that is cleaner, greener and sustainable, one that avoids the coming climate apocalypse, but they all try to do it without abandoning the idea of growth and, thus, forcing living standards into decline. The goal keeps being to maintain our quality of life, and keep everyone on the planet alive and happy, healthy and consuming.

We are literally the stereotypical lung cancer patient who keeps sucking down two packs a day while in treatment, or the alcoholic trying to get a liver transplant just to keep drinking copiously. It just won't work. And as a goal, it is counterproductive.

Slowly, a lot more people are beginning to comprehend what is at stake here. That if we carry on growing the global economy at its current rate, while at the same time relying on fossil fuels to power that growth, the planet is going to roast, and us with it. Not everybody believes this though, not by far. Those of us who do are still quite the minority. One of the roadblocks faced by those who wish to reduce fossil fuel use is that there is no consensus on how to tackle climate change. There is very little consensus on anything these days, really. And without a clear majority speaking together, there is no political will to do anything. Take it a step further, into the international or business realm, and the division only grows exponentially. And no matter what, the business-as-usual camp continues to claim that the scientific evidence is wrong about climate change, or that the scientists themselves have exaggerated the risks at the behest of opposing parties. Their answer to the question of what if it is true is simply that such problems can be delt with if and when they actually happen.

For most nations, fossil fuels are an existential issue. They either only maintain their current status because of the reserves they have to supply the world with, or else they are still developing and rely on these fuels to power their growing economies. Most nations of the Middle East certainly don't want to see fossil fuels phased out. Venezuela has the world's largest oil reserves…and not much else. Major powers like India and China exist as powers solely because of their vast use of fossil fuels to drive industry and manufacture, and it is hard to see Vladimir Putin being too interested in a reduction plan. This doesn't even get into the sheer mass of smaller nations struggling every day just to claw for survival, and that survival is inextricably linked to fossil fuels.

But the idea is now out there, and slowly growing, because some policymakers and industrial titans have now woken up to the risks of climate change. Risk that are growing by the day. There have been two terrible realizations by some few out there. First, that we have waited too long to get started, and it is too late now. And the second is that it doesn't matter how much solar and wind power gets put in place because we will still be burning all the coal, oil and gas we can because demand is only on the rise. Even if we burn it at a slower pace, it will still end up in the atmosphere and cause climate change. Right this minute, we are not feeling the effects of today's emissions, because there is a delay. The havoc being wreaked on the world right now, this is from the emissions of a long time ago. Stop emitting now, and we keep right on warming.

The time to do something was in the past quarter of a century, or better yet 50 years ago, when nations could have been putting together the infrastructure and policy for a new green economy and sustainable energy. Instead, we have all been going as fast as we can in the opposite direction. We have invested in fossil fuel-burning power plants and built energy-inefficient buildings in cities designed around cars. We have based the global economy almost entirely around the petrodollar, and we have created supply chains that stretch across the globe requiring massive amounts of fossil fueled transport and processing to get everything where it needs to go just-in-time. Our world may seem to run smoothly, but it only does so because it runs on carbon.

Climate change due to fossil fuel use is the existential issue of humanity, but the problem is literally that it is too big and too far reaching. Everything depends on fossil fuels, and the consequences of any reduction will be felt immediately, but the risks are weighted to the future, sometimes the far future. Banks, corporations, and national treasuries are set up to deal with short-term problems, such as an increase in inflation, current unemployment, and above all things, next quarters revenue and profits. As for the few governments that are thinking about climate change, they have other more immediate priorities: maintaining economic growth, increasing living standards, fending off threats to national security, and above everything else, getting re-elected.

Now, here is the rub. These two parties don't really answer to the people anywhere. They answer to each other. The politician will not get re-elected without the money and backing of the biggest of businesses, and those businesses will not get their revenues and profits without the power of politicians to make favorable fiscal policy.

The current crisis and conflict the world finds itself in right now is a great example of this symbiotic relationship at work.

On the one hand, there is the dire climate crisis presented most recently by the IPCC. They basically said, and I am paraphrasing here, "We are fucked if emissions from burning fossil fuels do not peak by 2025, and then get cut in half by 2030. The planet will cook us and all of the Beanie Babies and Pokémon cards we have been collecting will be worthless!" Something like that.

On the other hand, we have a world at war, not just among major energy producers but a world food breadbasket as well. That war, along with other factors, has triggered an extreme shortage of fossil fuels where they are needed, and a hefty rise in their cost as well. That increases the cost of everything, because remember, fossil fuels are tied into every aspect of modern life. People are about to be hungry, broke, and freezing, and they are pissed off. The economy is teetering on the edge of the worst crash and depression the world has ever seen, and that could happen at any moment. As I write this, it is July 29th of 2022, and so it is quite possible that it might have started before you even got this book.

In fact, yesterday, July 28th of 2022 was Earth Overshoot Day. In case you didn't know, Earth Overshoot Day marks the date when humanity has used all the biological resources that Earth regenerates during the entire year. Back in 1971, when they first started keeping track, we made it about even. Overshoot day didn't fall until December 25th that year, so we only overshot the mark by a few days. But now, we are almost in the middle of the year, and we have already used up our budget of resources. We are right now basically using up the resources of not just one Earth, but almost an entire other planet Earth. To meet the demands of today, we have to produce the resources of 1.75 Earths. That's a pretty serious overshoot, don't you think?

But back to the issue of climate crisis vs. energy demand. What exactly have we done about it right now? Which of these two sides of the issue are we addressing? Well, let's look at the recent actions taken, shall we?

The war Ukraine has caused a new "gold rush" for fossil fuel projects. Everywhere you turn there is a new media story about some country restarting old coal power plants or increasing production of oil and gas to meet rising demand, most of Europe is scrambling to find any and all replacements for all the fuel they used to get from Russia. Meanwhile, Russia is actually making more money selling less oil than before the war, due to the skyrocketing prices, and that is still with selling at a discount to nations like India and China which will take everything they can get their hands on right now. In America, President Biden is dumping out the nation's Strategic Reserves as fast as he can to try and lower gas prices even just a tiny bit in an effort to save his political party come the approaching elections, although turns out a good portion of those reserves are being sent to foreign nations, which is a real head-scratcher.

Soaring energy prices in the wake of Russia's invasion, rather than spark a move away from fossil fuels, have actually caused more investment in oil and gas projects, threatening to lock us into irreversible and catastrophic global warming. There are a huge number of new gas and oil projects starting, most of which won't even be built in time to combat the current energy crisis. They will, however, increase emissions in the long term and lock us into carbon-intensive infrastructure for decades to come. The world tossed aside the massive opportunity it had to use the post-pandemic recovery packages to support decarbonization of their economies, and now it seems like another rush for fossil fuel growth will happen all over again, due to our reaction to this new crisis.

Ignored is the latest IPCC report which warned that it is "now or never" for reducing greenhouse gas emissions. Never it is, or so it would seem. Everything going on shows that countries are instead investing even more in fossil fuels at this crucial time. Nothing has changed: we still go on responding to short term shocks like pandemics or conflicts by taking steps that will only increase emissions, totally ignoring the looming crisis of climate change in favor of fixing whatever the issue of the day is right now.

With these plans for more fossil fuel projects going ahead, they will end up as massively impacting our future fossil fuel use and lock the world into irreversible warming. And the crazy thing is that this goes completely against the clear agreement all of these nations came to with the Glasgow Pact made at COP26 not even a year ago in November of 2021. All of these world leaders and policymakers sat down and came up with a set of climate targets, and they gave great speeches about how dedicated they were to these emissions targets, bringing hope to so many…and then they threw them right in the trash at the first sign of trouble.

That's politicians for you. Whichever way the wind blows.

Carbon Bombing the Globe

So, we see what actions the political leaders have taken. But what about the business leaders? Well, while the worlds superpower leaders try to avoid dropping nuclear bombs on each other fighting over resources, the business sector has been getting some bombs of their own ready. These are "Carbon Bombs," and they are all set to usher in an apocalyptic climate breakdown. The industries of the world are waging a conflict of their own, a war directly against the planet itself.

What fresh hell is this? Let me explain. Carbon bombs are the work of the big fossil fuel industries, and they refer to the plethora of vast oil and gas projects being planned that are going to shatter that pesky 1.5°C climate barrier. These big dogs of industry plan to continue cashing in as the world burns down around the rest of us.

An investigation conducted by The Guardian identified about 200 in-the-works oil and natural gas drilling projects all over the world that threaten to completely blow up any chance of saving the climate. These projects, just on their own, would pump enough greenhouse gases into the atmosphere to take the planet far past the Paris climate accord's paltry limit of 1.5 degrees Celsius of warming.

According to the article in The Guardian, "These firms are in effect placing multibillion-dollar bets against humanity halting global heating. Their huge investments in new fossil fuel production could pay off only if countries fail to rapidly slash carbon emissions, which scientists say is vital."

So, basically, the companies of Big Oil are placing bets that no major moves will be made to restrict fossil fuels at all, and so they will have the freedom to do what they wish. Now, these guys aren't stupid. Evil, perhaps, but not stupid. They would not make this kind of investment in new fossil fuel projects unless they had a good reason to believe that all that talk about fighting for emissions reduction was bullshit. Thus, we go back to those political leaders once again, and I'm thinking we can see pretty easily where most of the campaign money will be coming from.

Despite the IPCC report stating that any more delays in cutting fossil fuel use would mean missing our last chance "to secure a livable and sustainable future for all," these companies are going hard at the money. And a lot of money it is. Trillions of dollars, in fact. The fossil fuel industry's short-term expansion plans have already been started, over half of them are pumping right now, and the rush is on.

Usually, it is places like the Middle East and Russia that tend to spawn the most future oil and gas production projects, but in this case, it is mostly North America and Australia with the biggest tally of these carbon bomb plans for expansion. The reaction to Russia's war in Ukraine has pushed oil and gas prices even higher, further incentivizing bets on new fields in The US and Canada and infrastructure that would last decades.

I highly encourage you to read the article by The Guardian which details the results of their investigation. It is titled "Revealed: the 'carbon bombs' set to trigger catastrophic climate breakdown." Written by authors Damian Carrington and Matthew Taylor, it is a very in-depth piece that will open your eyes to the problem, and to the fact that there are no easy solutions.

The Real Energy War

Make no mistake about it, there is also a lesser known, but just as serious, conflict afoot on top of the other more easily viewed wars, and it is one waged by the fossil fuels industries directly against the people of the world. Carbon bombing is an accurate term, as this is indeed a bombardment, and one that began long ago. It is not as if the industry was misinformed or unaware of what they were doing. No, it has been a conscious action to wage this campaign of theirs.

Way back in the 1970s, Exxon was busy conducting a lot of climate change research, looking to find out what effect their activities had on the world. They discovered that there was an undeniable link between the wildly increasing CO2 emissions and climate change. Real, hard science. And that was when they had a choice to make. And make it they did, a deliberate decision to not only continue to exploit the world for profit, but to greatly expand their efforts. That's right. They knew about the link between fossil fuels and climate change About 50 years ago. The entire fossil fuel industry knew as well. And in the face of such information, they went to war on the public perceptions to enable them to continue doing what they were doing.

One of the things they started with was a massive public misinformation campaign that convinced people that climate change and global warming was a myth. They were the driving force behind climate denial, and the public ate it up because it was exactly what they wanted to hear. Very similar to today's hopium and greenwashing campaigns. They even managed to lobby and persuade the U.S. Congress, most of whom were already in their pockets anyway, and that is how the 1998 Kyoto Protocol got voted down. The Kyoto Protocol was an early international climate treaty, one that might have been an excellent start for the world to be on track for avoiding the disaster we are facing now. While some nations did eventually decide to participate, the fossil fuels industry campaign managed to convince the world's two largest greenhouse gas emitters to stay out of it, and that would be the U.S. and China. Keeping them out basically defanged the Kyoto Protocol, because really, if the big dogs don't go with it, what hope does the rest of the world have to make any difference?

There are plenty of people out there who will say this is just a conspiracy theory. Maybe so, but we do know that conspiracies really happen. The thing is, in the early days of them, people like you and me don't really have much information to go on. In fact, most of the time it turns out to be revealed later that the public was only given half-truths at best, and often even outright lies. Primarily this comes through the various media enterprises out there that are owned by the wealthy elite and their seemingly never-ending supply of corrupted political leaders. Incredible amounts of money changes hands on a regular basis, and deals are made that we the people never even know of. Just look at the mess surrounding Jeffery Epstein and Wesley Wexner. Even know, with all of that concluded for the most part, what more do we really know about what the hell was going on? That network involved everything from Presidents to Princes, and we still have not been told anything concrete about what the hell was actually going on.

However, in the case of the fossil fuels industry campaign against the planet, there are some real facts available to look at. With these facts and an idea about how human nature has worked throughout history, we can figure what is really going on, and with a high degree of certainty. Climate change is indisputably real, and it represents a proven existential threat to our very survival. That is a scientific fact. It is also a fact that this climate change is anthropomorphic, meaning it is caused by human activity in emitting greenhouse gasses. By looking at the records of the big oil and gas companies, we can see that they not only knew this long ago but that they covered it up intentionally. There has been no real coordinated political action in the world to reduce the threat, and the fossil fuel industry has been allowed to run full steam ahead, aided by blatant corruption in political systems worldwide. Even now, as COP26 was taking place, they were already going ahead with investment into new oil and gas extraction, almost as if they knew in advance that nothing would be done to interfere with their "carbon bombs."

Given these things as facts, we can clearly see who benefits from the continuance of this Business-As-Usual approach to fossil fuels. And finding out who benefits, provides us with the motive that identifies a perpetrator.

And so. We know that these actions can only lead to our demise. There is no other possibility. Knowing that, why would these people choose that path?

Because they can. Simple as that. They have enormous power and wealth already, more than enough to prepare themselves to rule over whatever fractured bits of civilization remains down the road in a post-collapse world. They know that what they have done in the past has already doomed civilization to failure. Just as they knew about the link between fossil fuels and climate change so long ago, they too know about what is coming for us next. And again, they are continuing to go forward. And they know they can get away with it. What happens to the planet is of no concern to them, only that they continue to be on the top of the heap for whatever remains of it, and their lifetimes are not that much longer. Furthermore, they are not even worried. What can anyone do to stop them?

    What are we going to do? Rise up? How? And do what exactly? We are all exhausted by the survival mode of living that they have put us in, we don't have the energy to rise up. And even if some of us could scape together some ability to manage that feat, it wouldn't even be close to everyone. Most people still don't even see the problem because they have been brainwashed to believe that this society we have now is all there is for us. Even if that small percentage who know the truth were to rise up, it wouldn't matter. Their numbers are too small, and the police forces controlled by the corrupted government would simply lock them away. But they don't even have to go that far, because even those who know are still conditioned to believe their small actions can matter. Their peaceful protests, their boycotts, and all that. The Fossil fuels industry laughs at all of it, because they are the ones who put those ideas out there for us to devote ourselves to. All it is part of the same bug lie, and it is one that has been working for decade upon decade of peaceful "action." Just vote harder, people, that'll show 'em!

    The truth is that the Fossil Fuels industry, as a collective, totally owns and controls the media, the governments, and even us. Gas prices are at record highs, and we are still driving to work! We are still paying it! Even as the companies of Big Oil put out releases declaring record high profits, we are still in compliance.

    Control that they don't even have to try and enforce is the best control of all. We have all become our own handlers.

    Conspiracy theory? Maybe. But ask yourself a question sometime, when you are filling up you tank. When you are paying ultra-high prices for something you do not want, in order to drive to a place you do not want to go, and do something you do not want to do, just ask yourself why. Really think about it. Try and shed all of the normal excuses, such as saying that you have to do it or else you will lose everything. Is that really the reason? And if that is the case, doesn't that sound a little like slavery if you truly do not have a viable choice in the matter?

Either way, whatever you come up with for yourself personally, know that the vast majority of people will bow to this as necessity. And that is the result of the war that has been waged upon us by these industries for the entirety of our lives. And that is why they will win in the end.

The Wars Within Us: Civil Divide

Another very important entry into the conflict category is the growing swell of civil unrest in the world, and all the causes for it. While we may not rise against the powers that be in the world, we are certainly capable of rising against each other. There is a conflict within each and every one of us, right now.

There is hunger, poverty, and fear everywhere. The incredible spread of misinformation, and the fact that it is actually very hard for people to even identify what is and is not true. From blatant propaganda, media narratives, and outright conspiracy theories that even sound a bit plausible given the state of the world and all the strange stuff happening constantly. All of this, and more, is contributing to the growth of stark political, social, and cultural divisions among people everywhere.

Furthermore, these elements of people's general unease are being taken advantage of left and right. There are a great many forces doing so, from political parties galvanizing their side against the other, to those with racial hatred seeking to radicalize others to their cause, and even religious entities looking to whip their flocks into a fervor of anger that they can then use to increase their own power and relevance. Everywhere we look, we are being divided, red and blue, left and right, Coke or Pepsi...

Never before have people been so torn apart by so many issues all at once. Race, gender, religion, those have always been some of the big ones. But now, every little thing is a big deal. And all of those little things are being used as weapons by various parties to drive us all further apart. A lot has been said lately about the potential for the United States to "Balkanize," or separate into a bunch of separate smaller countries, but it is more than that. This is about the balkanization of all of humanity. We are reverting to stark tribalism rapidly, and if that is not an indicator of societal collapse, I don't know what is. The entire concept of society itself cannot work without the vast majority of people being able to live and work together toward a common peace and stability.

What is threatened by this is our very social cohesion as a people. And without that glue to hold our nations together, the world just comes apart.

The Rise of Fear, Uncertainty, and Doubt

The fact that violence is on the rise everywhere is not something I should have to convince you of. You are seeing it everywhere around you, pretty much every day. The big ones are obvious, but what may be less obvious are the little instances that are occurring with rapidly increasing frequency. There is something there, something that we are all feeling whether we are consciously aware of it or not, and it is causing a general decline in our relations.

There are many contributing factors and symptoms of this decline. From the growing political divide and increasing polarization of people's beliefs, to basic stresses and fears with regards to wealth inequality, security, and even our very futures.

In the financial world of investing, I first learned of a thing called "FUD." FUD stands for "Fear, Uncertainty, and Doubt." It is often a clever tactic used to misdirect people into being afraid or uncertain of an investment and doubt the odds of such being successful. An example would be a story put out about something like a certain company which casts some negative light on that company's coming odds of success. Maybe there are rumors about a crippling lawsuit, or embezzlement on the part of a corporate director. What it actually is doesn't matter. What matters is that people get wind of this story, and immediately they become stricken with FUD. They begin to worry that they may lose money from holding stock in the company, or whatever. The point is that the feelings imparted cause a reaction that they would not normally have taken. Selling off their stock or investing in a competitor instead. A strong enough FUD can crash an entire market. Just look at what happened to the cryptocurrency markets every time China threatened to ban its use and mining. Large drops in market share and valuation as people began panic-selling, followed by steep rises again once the FUD was realized.

Now, what I am talking about here has nothing to do with the normal definitions of FUD. Normally, all FUD has a target. A company, a market, even a political candidate or a nations currency. What we are experiencing across the globe is a new kind of FUD. This one is undirected. There is no target for it, nor is there any singer perpetrator consciously putting it out on purpose to bring about a certain effect. This FUD was not put out there by anyone, and it has no definable relevancy to any specific thing. It has just become manifested in our lives.

Think about it. What if a friend proposed a trip to a nearby town right now, maybe to go see your favorite band in concert? How does your mind react immediately? Well, at one time, there probably would have been a rush of excitement at this cool idea, an eagerness to get it rolling, and an anticipation of all the fun that would be had. It sounds awesome, let's go!

But now? Now your mind starts to flash in different directions. It is so far, and these gas prices are crazy. It seems expensive, you need to make sure you have the money to pay the bills this month. A concert? Man, what about the danger of all these shootings and attacks lately… Fear, Uncertainty, Doubt.

That is the FUD I am talking about. And it is something that is weighing heavily on people. And not just here and there, no, it is part of everything. People question everything now, even going out to get a cup of coffee, or whether to upgrade their old sofa, even with dating everyone is on edge about something being put in their drink or finding themselves being taken advantage of. This FUD is pervasive and ever-present. Everyone is feeling it all the time, some harder than others, and it is taking its toll on our mental and emotional health.

It does a great deal of damage to feel even a small fear if you have to feel it constantly. And such feeling can build and feed off of all of the other experiences in your life, changing them, heightening them, until what once would have been a minor annoyance to be shrugged off blows up into a full manifestation of rage and anger that has no real target.

It has been building within society for a very long time, but only recently has it gone into overdrive. It really became noticeable with the onset of the COVID-19 pandemic. That was a real fear right there, a true feeling of uncertainty, and even in many cases a doubt that we would survive. It hit our collective psyche hard. And it isn't even over yet, but all of a sudden, things began to happen at an accelerated rate. Everything from issues with poor wages, political divides, wealth inequality, market instability, inflation, the climate emergency, food insecurity, war… there is no need to continue with the entire list, it is way to long for that. The point is that after the pandemic hit, the hits just kept on coming, and they have not let up, not once. In fact, they are accelerating.

All of this has led to the rise within people of an all-encompassing FUD for every aspect of life. Everything is bleak and growing bleaker, and the stress continues to build. And then, it boils over. People lose it. Sometimes just for a moment, throwing coffee at another person for some slight, real or perceived, or angrily cutting off another driver in traffic and shouting insults in passing. But other times, the snap is much more serious, and much deeper, when it comes. People start shooting up schools or lighting people on fire…

Everyone is under a feeling of a heavy, pervasive, and unrelenting weight of contradictory emotions at all times, and no one really knows where to direct it. Because of that, we are all slowly "going postal" on a societal scale.

Discontent with the World

We can call this phenomenon Mass Societal Discontent, which I will shorten here to MSD because you don't want to keep reading it, and I don't want to type it over and over. This is just a term I created to put a label on what I am observing, and while I am sure there is an appropriate scientific term for it, I don't know what it is. This is a layman's guide, after all, so MSD it is.

So, what exactly is MSD? It is an unconscious and pervasive worry among people about the decay and instability of society, which manifests as an unmanageable deterioration of their emotional state. There are 5 general aspects of this manifestation to be seen at the large scale: loss of trust between people, confusion of ideals, the decay of truth, breakdown of the community bond, and rising instances of violence.

These all swirl together with each other. Loss of trust in someone leads to the decay of truth, because whether what they are saying is true or not, we simply do not believe them anymore. We want to double check and "do our own research," but where is the truth in this vast sea of conflicting information that we have instantly ready at our fingertips? We don't know, because we don't even trust those sources of information anymore. And so, we have a general decay of truth, of what actually is the factual state of a thing. We can look and look and look, but we just can't seem to find a rock-solid source that we can believe.

The loss of trust we once had for our governments, our media journalists, and even each other, is profound. Never before has distrust been so rampant. No one trusts anything anymore, and reinforcing that is the fact that for any "truth" to be found there is quick and instant access to an opposing opinion, scientific study, or religious ideal that shows it to be false. We don't believe each other about anything anymore. No subject is safe from controversy. Was the moon landing real? Could the earth really be flat? Will my 5G phone service improve if I get the COVID vaccine?

These may sound ridiculous to you (I hope), but these are all real divides among people today. Serious ones, too. And when we are confronted with people who do not share our exact worldview, any hope of trust just evaporates. MSD has made us question anything and anyone that does not line up perfectly with the image we have in our own heads about what is right.

Because of this we become confused as to what our own ideology really is. Are you a democrat or a republican? That was a pretty simple distinction once, but not anymore. Now, there are an uncountable number of divisions within them both, some even overlapping. It's almost like the difference in music. When I was younger, we had techno. That kinda morphed into EDM, or Electronic Dance Music, and then it branched into so many subcategories that I have no idea what I even like anymore. Is it Glitch Hop? Deep House? Vocal Trance or Liquid Funk? Do you even know what the hell I am talking about right now? Do I?

Our ideology is fractured on a million different lines, and where once those devotees of several could coexist under an all-encompassing banner, it is no longer like that. Now, we separate quickly into tight little camps of 100% agreement, and glower at everyone outside as if they were an enemy trying to trick us or take advantage of us.

Like Treebeard from The Lord of The Rings, when asked who's side he was on. He said, "Side? I am on nobody's side, because nobody is on my side."

And that is it. We are, all of us, slowly becoming beholden to "sides" in a free-for-all battle royale of our own making, where everyone stands, or falls, alone. And that inevitably leads to the breakdown of community bonds between us. We start distrusting and hating our neighbors, suspicious of their intentions and allegiances. We don't want to associate with people that do not agree with us, who may be working against us, and after a while that starts to seem like everyone.

A community cannot thrive without that bond. Without neighbors helping each other and enjoying each other's company. Without taking responsibility not just for ourselves, but for everyone around us, to ensure the safety of the whole rather than just the individual. Nations cannot survive without functional communities, and when nations start to face troubles, the entire world bears the consequences.

And the inevitable result of the increase in MSD is the eruption of violence. It began small enough. People become more irritable, less sociable. Fights are picked more often, and disputes between neighbors get more out of hand. There is a gradual increase in things like every day violent crime. Domestic abuse, assaults, armed robberies. Some incidents become larger and more widespread, such as terrorism from bombings and mass shootings. Events even get randomized, as people just seem to snap and lash out at whatever vulnerable population that is nearby, angry at the universe and not knowing who to attack to make it all stop.

And through all of this physical manifestation of MSD in people, there are others for whom their own manifestation leads them to try and direct others. To gain a following and use the power it represents to advance their own side of things. People become polarized along major lines, with fancy slogans and engineered acronyms to make them feel like part of a team, the right team. MAGA. BLM. Proud Boys. They have war cries and chants, "Slava Ukraini!" or "Let's Go Brandon!" Whatever, they don't really matter. What matters is the width and depth of the divide between them, and it is growing wider and deeper every day.

MSD affects us all, even at the highest levels of our political, religious, and business leaders, and when the people who lead us are just as fed up as we are with…whatever…then the violence gains the potential to grow across boundaries and borders. Those already in places of power begin to use that power in increasingly extreme ways. Scoring the points today means more than securing the future for tomorrow, and that leads to chaos.

Take the recent U.S. Supreme Court decisions. Within a span of a few days, the court both stripped states of their ability to regulate the carrying of firearms, granted them the ability to regulate abortion as they saw fit, and defanged an already toothless Environmental Protection Agency's ability to regulate greenhouse emissions on a federal level. Three of the absolute most divisive and deeply held beliefs, the right-to-life vs the right-to-choose, the pro-gun vs anti-gun movements, and the climate activists vs "those who are in favor of the jobs that the comet will bring." They dropped bombshells on them all. No matter what beliefs you hold yourself, dear reader, or which side of those issues you are on, something about one of them probably pissed you off. If not, then the response from the currently ruling party certainly did. And now that same Supreme Court is on track to allow states to gerrymander voting districts, restrict same-sex marriage once again, and who knows what else. It is absolute madness. But the worst thing about all of this is that it does not seem to be madness for everyone. In fact, we as a people are split on the issue. Some of us think all this is a great idea. And in this era of disinformation and lack of trust, who's to say they aren't correct?

Where Things Are heading

People everywhere have had some shocks lately. Look at the historic riots in the face of the police brutality against George Floyd? The Canadian trucker convoys? The insurrection at the U.S. Capitol Building on January 6th? To the rest of the world, look at the civil unrest and complete economic collapse in Sri Lanka, or the farmer revolts in the Netherlands at the governmental attempts to reduce fertilizer usage. Even though the beginning of the pandemic brought a brief respite to the world regarding incidents of social unrest, uprisings continued through the later stages of the pandemic, with events in both advanced nations and developing ones as well. In the former, protests erupted in places where major social unrest is usually pretty rare, often with anti-government or anti-lockdown motives, in the U.S. of course, but also including in Canada, New Zealand, Austria, and the Netherlands. In emerging markets and developing economies, the apparent motives for recent unrest have been more diverse, with examples including anti-government protests in Kazakhstan and Chad, and a coup in Burkina Faso, which was left out of your western media diet, I am sure. There were also regional protests in Tajikistan, a constitutional crisis among other crises in Sudan, and of course Ecuador looks like they want to try on another government yet again.

In coming months, several important factors could lead to an increased risk of future unrest rising even more steeply than it has been. For one, public frustration with rising food and fuel prices will no doubt increase. Those prices and shortages are only projected to get more severe, and although the economic causes of civil disorder are complex, and unrest is exceptionally hard to predict, steep price increases for food and fuel have been associated with more frequent civil unrest in the past. Just look at the Arab Spring.

According to what I have gathered from several indices and civil unrest projections, over 70 countries will likely experience an increase in civil unrest by the end of 2022. About half of these nations will be in Europe and the Americas and will likely see a particularly significant deterioration and rise in both protest and rioting. The western world is entering a tense summer right now, but as bad as things are we are still working off of our reserves and the expectation that things should soon return to "normal." But what happens when things get worse instead of better? When people's savings begin to run out, and unemployment begins to rise with inflation and gas prices? How will the "developed" world react to suddenly sliding towards a state normally reserved for other countries seen on TV?

Globally, the increases in violence and protests are expected to be driven mainly by food insecurity, energy shortages, and rampant inflation. But there are also other considerations, such as the rapid erosion of mechanisms that have historically helped defuse such tensions, like the freedoms of assembly and the press, and the rights of people to live their lives and speak their minds. It is one thing to contemplate possibly losing your job or getting arrested for speaking your mind, but without anything to lose, what will hold people back? I fully expect the surge in instability to take place against a backdrop of the painful post-pandemic economic disasters, along with the uncertainty of escalating war, and the growing loss of freedoms combined with rising costs of living, all of them things that will likely inflame existing public dissatisfaction with governments.

With no end to the pandemic-induced economic downturn coming soon, and the resulting shock of war raging in Europe, the surge in civil unrest is unlikely to subside any time soon. In fact, we may be living in the most peaceful time we have left. What goes up must come down, true, but that can take quite some time, and who knows what kind of condition the world will be in.

While this past couple year period from 2020 to 2022 has established itself as one of the most tumultuous periods in recent history, most projections show that the worst is most certainly yet to come for many countries, and indeed, for the entire world. In fact, I expect 2022 to be not an outlier but rather a harbinger of things to come. And not just within the next couple years, but for much of the coming decade. In the most nations the ranks of protesters and rioters marching against long-standing grievances will likely continue to grow, both in size and in violence, with millions of newly unemployed, underpaid, and underfed citizens, posing a risk to all domestic stability. And keep in mind, this will be happening with a war that is escalating and others about to start. And with the pressures of climate change bearing down on us all harder and harder every day. This is not like any other collapse scenario with historical parallels we can compare to. In the history of humanity, there has never been a major hot war between multiple nuclear armed opponents. There has never been a global financial collapse on the scale of what is possible today. And none of the previous human civilizations that have collapsed had to deal with a planetary biosphere that was dying along with them.

People, on an individual level, are becoming more unstable. They cannot all deal with the pressures they are facing in the same way, and few today are prepared to face such things. In the early days of the Industrial Revolution, suddenly losing it all and having to fall back societally, well, that would have sucked, but people could actually remember, in their own lives, what it was like before all of it. Losing engines and whatnot didn't mean nearly as much to them, and they had the confidence and the know-how to struggle through it. But it will matter a whole lot more to the average "developed" human of today who can't even find their way across town without a pocket-sized supercomputer and access to a world-wide information database. To face a descent from 2022 all the way to 1722 is just unfathomable, and insanity inducing. Not for the Amish, perhaps, or the poor villagers in remote corners of those nations that have not yet fully "developed," but the rest of us are screwed.

And that is the mentality being subconsciously manifested as MSD, or whatever the appropriate term might be. We can feel it, all of us. Something is wrong, something bad is coming, it is constantly there like a rat scratching at the attic walls inside the home of our minds. We just can't articulate what it is, or where it is coming from, but we certainly feel it.

Think of it like the phenomenon of "Infrasound." Infrasound, also called "the fear frequency" is a low frequency sound beyond that which can normally be heard by humans. But it does have vibrations, and science has been recently studying how these constant vibrations can wear on people, grate on their nerves and psyche and eventually even cause them to experience physical symptoms like hyperventilating and hallucinations. I am not going into all that, because that isn't the point. Go check it out for yourself.

But that is how I see this feeling that causes the MSD I have described. It is an ever-present and all-encompassing…unease. It creeps into your life slowly, like a stalking animal in the dark of night, and once it is with you, it probably never goes away. Too many things just going wrong, no positive outlook on the future or even the present, things just keep getting worse and worse…and there it is. Next thing you know, you are on the internet learning that you are a "doomer." If you don't already know what that is, don't worry, you will figure it out soon enough.

This overwhelming and pervading sense of doom we are all experiencing as this MSD, this is what is driving us all slowly over the edge. We are the first humans in history to be facing the entirety of the world's ills, all at once, available all the time instantly on our phones. We basically have a direct line from all the bad things happening in the world pouring right into our brains, and the algorithms of our apps and devices focus it constantly to show us more and more of what we are constantly viewing. A person can, and has, actually live-streamed the mass shooting of a bunch of people. Yeah, like a game on Twitch. Right now, you can flick open an app and view a live feed directly to some poor bastards under aerial bombardment by Russian warplanes in Ukraine or witness the dying breaths of sickened patients in a Sudanese hospital, or see countless and endless streams of pictures and video detailing every horror playing out across the world, right now in real time.

Not only is our world coming apart at the seams, but we are too, and we are watching it happen all the time. It is no wonder we can't get it out of our heads.

And that will be a big part of how we respond to the breaking of our world. We are on the edge, and the chasm below is deep, indeed.

The Decay of Truth

Sometime in the last decade or so, civil discourse in the world, particularly in America, has been suffering from a gradual decline in the ability to recognize and accept truth. There are multiple trends that demonstrate this, such as an increasing tendency for people to disagree about facts and analysis of factual data. Contributing to this is a blurring of the line between what is opinion and what is proven fact. Among regular people, politicians, media pundits and all the rest, there has also been a dramatic increase in the volume and influence of opinions and personal experience over previously known and accepted information. This has resulted in much lowered levels of trust in almost every source of information out there, whether proven or not.

These trends have a great many different causes, and I will not try to go into all of them in detail here. We have all seen it, and most of us have experienced or participated in it to some degree. One of the factors is the rise in cognitive bias, which has been influenced extensively by changes in the systems we use to gather information. Social media is now a large part of how we discover and share information, and often that information gets discussed and shared on forums where all the users are of a relatively similar opinion about a particular subject, and so they turn into echo chambers that only reinforce a limited view. The 24-hour news cycle we live with, and the instant access to every bit of it anytime and anywhere, is another contributor. These things are relatively new experiences for humanity, I am not sure we are adjusting very well. Many different competing non-curricula demands on the education system have generally diminished the amount of time spent on factual literacy and critical thinking skills as well. Finally, there is the increasing polarization of people, along both political lines and social groups. Rather than show information objectively, everything that is presented to us, by journalists, teachers, and politicians is always accompanied by its most divisive counter argument as being the "wrong" side of the issue.

There are many problems with all of this, but probably the most damaging consequences of this decay of what the truth really is, societally speaking, is the division it is creating. We are becoming so divided that it is almost as if we expect everyone to agree with us 100%, and even the slightest deviation on a single issue can get another person labelled as "the enemy." There has been a complete erosion of civil discourse among people, and the lack of firm agreement on issues has led to political paralysis among leaders worried about stepping outside the lines of too many voters. Furthermore, all of this disagreement leads to alienation and disengagement of individuals from political and social systems that we depend on.

No More Trust

With the uncertainty of what the truth is, we also end up with a general decline in trust, both between the individual people and even institutions and governments. As the loss of shared trust intensifies between people, they start to feel disgusted by the state of society, and especially with what they see as the inappropriate reactions of other people around them. Trust in institutions and governments plummets. Moral indignation becomes widespread. And contempt for the established Powers-That-Be grows ever more intense.

People begin separating into tighter and tighter groups, almost tribally. Highly moralistic leaders appear on the scene, very uncompromising ones, and they reinforce the belief that their issue is THE issue. These groups will use all the new modes of communication to hijack control of the national conversation wherever they can. Talk begins to grow that they should "rise up" and take control over the system that they see as failing the world and change it to a system that will address their issue primarily. These are times of agitation and excitement, frenzy and accusation, hysteria and anger. And times like these usually lead to great changes, and not always for the better. Balkanization of the United States? A civil war within our borders has been talked about more and more of late, but rather than there just being two sides with two positions we have hundreds of sides with differing positions. That is chaos, not war.

Our own civilization-wide convulsion of anger and distrust began somewhat recently, about the mid-2010s, with the rise of a range of groups pushing various issues. I won't be going into them all here, you know the stuff I am talking about. Tribes began springing up everywhere, centered around all sorts of things, some good and some not, but either way they were all extremely distrustful of each other. More recently, the events of 2020 began bringing the chaos to the fore. The coronavirus pandemic and subsequent battle of vaccine science and control politics; the murder of George Floyd at the hands of the police, and the worldwide protests and riots that spawned; militias recruiting and racial hate groups coming out of the woodwork with social-media mobs, and urban unrest…those things were like bombs that dropped in the middle of the hurricane that human societal interaction had become. They did not necessarily cause the increase in distrust, but they accelerated every trend and exposed every flaw in our global society.

Now, as we enter the second half of 2022, this convulsion of distrust and division is rapidly rising toward a climax. Political leaders in the world are in the process of shredding every norm of decent behavior and wrecking every tradition they touch. Regular people, once functioning fine in society are now unable to behave responsibly, unable to protect themselves from the actions of their leaders, unable to even discern the truth about what is going on in the world. They have no idea where to turn, and so they are turning on each other. In the middle between the people and the governments, we have the media undermining the basic credibility of the news and information we receive, and arousing suspicions that every word and act that differs from the accepted opinion or party line is a lie and a fraud.

All of this, along with everything else, seriously threatens to undermine the functioning of our national governments and incite vicious conflagrations that could leave the world covered in the remains of charred and shattered nations. Distrust is the final instrument of this crisis, but the conditions that led to it and make it so dangerous at this moment were decades in the making, and those conditions will not disappear anytime soon.

The Crisis of Truth and Trust

What does it mean for us when we cannot discern the truth or trust others to reveal it to us? Trust and truth are both eroding in the United States, and indeed in the entire world, and at a time when we are entering a period of catastrophe. There are economic crises emerging, the effects of climate change are starting to hit home, and the world is embroiled in conflict, amongst many other things. And to add to that, we have this crisis of reality. It has gotten much harder to figure out what is and isn't true, to identify the boundaries are between fact, opinion, and disinformation. And that can be something of a "force multiplier" for all the other issues we are facing.

The lifeblood of civilization is a common understanding of the facts and information that we can collectively use as the basis for negotiation and compromise. When that goes away, and it most certainly has now, it can shake the whole foundation of civilization. This is a global issue, not just an American one, although we do tend to make a grand display of ours. But all across the globe, ignorance of the truth and lack of trust in those who provide it is undermining confidence in all of our institutions. If we do not have an agreed upon set of facts on which to base decisions, then it's really hard to judge whether something is good or bad, such as a company or a politician, or even position on an issue. We are rapidly losing the ability to make good decisions, or even any decisions at all. It leads to a kind of paralysis, both on the part of leaders and also regular people. No one can gain a concurrence on the issues, and thus actions are not taken simply because we cannot agree on what to do. When the people in a civilization lose belief and trust in their institutions and in each other, that civilization collapses.

And that is what we are looking at now. A multi-headed hydra of emerging leaders with many different ideas about what to do, each with their own followings that are vehemently opposed to any actions suggested by others. The various politicians are choosing sides, the media outlets are choosing sides, and all of the populations are also choosing sides, but no one seems to realize that, with the very planet at stake, there is really only one side we all need to be on.

The MSD we talked about is a symptom of this disease. Everyone is feeling it, some more than others, and it is driving the rifts between people farther and farther apart. That has left us a broken, alienated society caught in a distrust feedback loop. It has given rise to everything from conspiracy theories and religious cult movements to the increasingly common attacks based on racial, political, and cultural lines. Soon, it will give us complete chaos and the breakdown of all social order into a lawless wasteland of our own making.

6

THE PEAK OF COMPLEXITY

The Fourth C: Complexity

Human civilization is highly complex, interconnected and globally interdependent. The centuries following the Industrial Revolution has given us the ability to exploit large energy resources, primarily the fossil fuel types, and we used it to greatly boost growth in both industrial and economic output, along with the massive increase to global population that came with them. The agricultural revolution of the mid-20th century spawned even more acceleration to our growth through increased food production. This growth model was made the very foundation for our modern civilization, and a steady increase in economic growth became the standard against which all things were judged. That growth was accompanied by a substantial and ongoing increase in societal complexity across every facet of civilization. However, it also resulted in a strong and increasing amount of pressure on the Earth system that we all depend on to survive. A pressure that was long ignored in favor of increased growth and short-term utility.

For many years, researchers and scientists have explored the growing environmental impact of human expansion and the speed at which we have been exploiting available resources. As we have already gone over, the 1972 MIT study that came to be known as "The Limits to Growth" was undertaken to delve into the broad dynamics of global systems and their relationship to the human activities of resource depletion, land use, and environmental pollution. The scenarios for the future that the study revealed were grim.

The most likely conclusion was that civilization was headed for overshoot and collapse as the result of the current growth trends in world population, industrialization, agricultural expansion, and resource depletion. The researchers warned that if this trend continued, we would soon reach the limits to the Earth's carrying capacity. Overshoot in this context means that society would soon reach the peak of that capacity, and then launch fully past it. Very quickly we would find that everything was running out, and as the reality crashed home, we would suddenly find ourselves without enough energy and food to sustain our current population and civilization, and that would lead to a steep drop-off into rapid collapse.

We have discussed a few of the studies that have attempted to figure out what these physical limits were that we must avoid overshooting. Of particular importance is the framework of planetary boundaries which defines nine specific areas of concern, and quantifies the safe operating parameters that, if crossed, could trigger an abrupt environmental change within our planetary operating systems. Why is this one so important? Remember, five of these nine boundaries have already been crossed. Accordingly, we are already beginning the overshoot and about to face catastrophic consequences for civilization.

Our collective exploitation of finite resources and the wanton destruction of ecosystems has led many modern researchers to conclude that we are about to face the dire consequences of those actions. We have discussed a few of them here, but there are a great many others. Remember the recent study we went over by Mauro Bologna and Gerardo Aquino? In closing that work, they concluded that the probability of our civilization surviving all that it currently faces was "less than 10% in the most optimistic scenario." Most. Optimistic. Scenario. Let that sink in again.

We are staking everything, including our very existence, on the continuance of industrial and technological complexity. But when that system stops working, we are going to lose it all.

A Vast and Tangled Web

No one really knows too much about the complex systems that we have collectively constructed to keep our technologically dependent society humming along smoothly. It's just not something we think about. There isn't really any reason to, and we never have time to waste these days, right? Gotta wake up, get that breakfast down and put in the coffee order to pick up on the way to work! Lots of things to do today, and deadlines are closing in. Your Apple Watch keeps you moving, reminds you to buy that gift for your niece's birthday party tonight, same-day shipping from Amazon, so you are all good there. Check the email for the reports you must present at the Zoom meeting later, the boss will be attending afterhours from Beijing, so you have to be sharp! There's also the notification from your smart home fridge that the milk has expired, but no worries, Alexa has that on the list already, letting you attend to more important things while your Tesla handles most of the work of driving downtown...

Who really considers the massive infrastructures that support all of that? But all of those everyday normal actions are made possible by an insanely complex, algorithmically calibrated, computer automated, and virtual slave-labor-dependent global network of systems and supply chains.

There are currently millions of shipping containers in circulation globally. At any given moment, about 5 million of them are crossing the sea somewhere. America alone imports something like 20 million massive containers worth of products every year. Everyone thinks about it in terms of smartphones and fancy sneakers when we consider imports from Asia, and sure, that's where they come from. But the real extent of the movement of goods is rarely seen. Most of those shipping containers are stuffed with almost everything else we interact with on a daily basis. Cups, umbrellas, markers, notepads, packing tape, building materials, blankets, car parts, frozen food, drugs, paint, light bulbs...everything. All of the endless lists of goods and materials that make our busy modern lives possible.

Intimately linked to that sprawling supply chain, is another vast and complex system which is mostly invisible to us. It is the international financial markets. This highly technology-driven network links together all of our banks, governments, stock markets, exchanges, news services, and hundreds of millions of individual users, both the human type and the automated computer ones. This system has evolved to a level of complexity that makes it completely unfathomable by any single person, and everything happening within it moves at a speed and scale that is far too great for a human to manage.

Average daily trading volumes on a large exchange usually surpasses billions and billions of shares, just in a single trading day. Buying, selling, moving, trading, shorting...everything moving at breakneck speeds. Hundreds of billions of dollars in value are traded every single day, just on a single exchange. The only way to manage a market of this size, scope, and complexity is a complete and total adoption of computer automation where we have basically handed over day-to-day decision-making to software programs. The industry of finance is obsessed entirely with growth, and so these systems have led to an exponential increase in complexity over past systems, and they get better and more complex by the day. Human traders might normally average a half dozen trades in a day, but the high-frequency trading algorithms can make thousands of trades every second of that day. No human can compete against that, and traders have mostly become nothing more than the caretakers of the systems that run their empires.

But we still dip our toes in the water here and there. Now, anyone can be a trader. There are the apps on our phones that automatically round up our debit card purchases and invest that spare change into fractional shares of companies and industries. We might have a general idea or a plan we put in place for what we want that money invested in, but the actual nitty-gritty of the matter has been handed off to a program. But we can get into it, if we want. Driving down the road, one could hear a comment made in a podcast, or see one of Elon Musk's infamous tweets, and instantly whip out their phone and make a trade on the New York Stock Exchange to buy, sell, or option whatever stock was mentioned, and do it in less time than it takes for the red light to turn green. Swipe, swipe, and done.

The technology behind the software platforms and algorithmic systems that cements all of these networks together has become an incredibly complex system itself. The internet is probably the system you are most familiar with, one which we interact with the most. We are intimately ensnared by it, and our everyday lives depend on it, but most of us give little thought about what lies behind the cracked touchscreens of our phones. The internet is a thing truly understood by only a very few. Made up of billions of users and their connected devices, from the size of massive data centers down to little old you and me swiping our way across the landscape of Tinder, the internet is a vast network dedicated to storing, creating, viewing, and moving data on a scale far beyond our individual comprehension.

Let's take a look at some common platforms for various media. YouTube is a great example. More than 500 hours of video is uploaded by users to the network every single minute, which is over 80 years' worth of video put up every day. And what about Facebook? Something like two-and-a-half billion Facebook users, with most of them logging on daily, posting, relying, checking in and uploading pictures. What about Reddit, and all the active subreddits that make up the forums? Hundreds of millions of users there, either spreading information, ranting, or just shitposting. Take a look at Twitter with over 500 million tweets being sent out each day, over 6,000 tweets every single second. Social media alone is incredibly complex and overwhelming to even think about. The sheer mass of content created is hundreds of millions of terabytes of data per day. That is mind boggling, and that is every single day and growing.

Humans have always sought for better. Bigger, better, faster, more. That is what growth and achievement is all about, right? We have made great strides in building this civilization of ours, but our growth entails costs in the form of complexity. To solve what we see as "problems" we create ever-more laws and regulations, with ever-increasing massive bureaucracies to manage them. We have gone from times when we pretty much relied on ourselves individually and at the small community level, to having an almost worldwide government to manage even the smallest of affairs. Can you drive a car? Once, that just meant whether or not you could operate the machine, but now it has to do with receiving the proper education in a myriad of traffic and regulatory laws, possessing insurance, getting specially licensed, and a whole host of complex processes that have very little to do with whether or not you simply know how to operate the machine at all. In fact, the actual operation is the most neglected part of the process!

But it is not just the complexity of bureaucratic processes. We have created untold numbers of new financial products, large data processing networks, and a vast range of companies, institutions, interconnections, structures, programs, products, and technologies. We often solve what we see as problems by creating entirely new roles for people. Bandits are a problem? Well, let's create a Sheriff to tackle that issue. To relieve us of the "problem" of dealing with it ourselves individually. We have created new institutions, like the United Nations, and developed new technologies to solve other problems, like our inventions of smartphones, ICBMs, and industrial agriculture, so that now we can have instant communication, wage war across vast distances, and feed an increasing population of people who are living longer and longer. We have advanced medicine in such a way as to almost completely remove the natural process of death by "natural causes" and as such we have millions of people that simply cannot even survive without the solutions offered by modern medicine. Does that sound like an ecosystem that nature designed to function successfully?

All of these things and more have added many, many level of complexity to our modern lives. And we accept or even demand this added complexity because we believe that there are benefits to solving these problems which outweigh any potential consequences. And there most certainly are benefits, at least if we evaluate things on a case-by-case basis. Anyone dependent on a rare synthetic drug would say it is definitely worth it, as would their families. Taken as whole, however, the unrelenting addition of ever-increasing complexity to civilization weighs heavily on the system, bogs it down, increases energy requirements, and will eventually outstrip available energy and food resources, setting the stage for collapse.

But we don't care about that. We have built and grown and discovered and exploited everything in the quest for "better." What we have ended up with is an interconnected global civilization built on the constant flow of capital, materials, and data, all of it available not sooner or later than needed, but "just in time." These just-in-time approaches to supply chain operations, and the financial transactions that go along with them, have spread to dominate global manufacturing in every aspect. And the computer networks we've built to manage these flows in the most cost-effective ways have become so complex that they're now beyond the scale of humans understanding them.

One can imagine these networks as if they were living organisms, massive ones, with webs of tentacles spanning the globe that touch everything and interlink with each other. But I tend to think of it more like a hive-mind collective of organisms, without any form of centralized influence. Like a nervous system without a brain for making decisions, functioning more like a human body on life support; brain dead but still each part of the body carrying on its autonomous functions. Because the reality is that these networks are not all being directed singularly. In fact, most of the parts are completely unaware of what the other parts may be doing at any given time.

It's kind of like a form of distributed intelligence or knowledge, where many different nodes of the system have only the limited information of their immediate environment. This decentralization of decision-making means that each node always does what it is supposed to do, right when it is supposed to do it, and it doesn't need to know what the other nodes are doing because they will all be doing everything exactly the same. And so, as long as things are running smoothly, they all interact in ways that lead to a collective decision-making without them even knowing individually that's what they are doing. The body can continue to function just fine without a brain. In fact, it is much more efficient that way.

But what happens to such a system when things are suddenly disrupted? The system functions well but requires smooth running across the board in order to do so. In this age of the so-called "4th industrial revolution" our most successful industrial, financial, and technological organizations are all complex and interdependent systems, each operating with very little redundancy between their various parts. The smallest of inputs can, and will, create major disruption at a scale almost impossible to predict. A minor shock applied to, or suffered at, a precise moment can overload an entire system. The major disruption that results is experienced in the form of events such as data loss, system outages, environmental damage and even loss of life accidents, just to mention a few.

We have seen some examples of how a small hiccup can destabilize the entire system. The blockage of the Suez Canal by a single ship caused a backup of hundreds of ships full of goods that were supposed to arrive somewhere "just in time." Since they did not arrive, other manufacturers did not have the materials to continue work for their own orders, and then finished goods could not be sent out to retailers, and so on down the line. This is because, while the system functions well, it can only do so as long as there are no problems. There is very little redundancy in place to handle problems. Literally, one little wrench in the gears, and the entire global machine grinds to a halt.

So, we have all these systems, all linked together and controlled by various types of algorithmic automation and computer logic frameworks. No one really knows at any given time why the system would tell a trader to take a specific position on a matter, or message a ship's captain to slow down a massive container ship, or any number of millions of other decisions that are being decided for us. We don't really know why, we just complete the action and assume that is for the best, because there is simply too much to know. And besides, that's why we made these systems, right? To have the capacity to know things about incredibly complex processes that are impossible for a single human to figure out fast enough to be of use.

This is fine, right? I mean, there are plenty of people, such as executives, financial analysts, and tech developers who have told us that it's better than fine; it's actually great. We should all be very happy about this. There is no way we could have the kinds of instant-gratification lives we do without such a system. No matter what, there will always be the right amounts of goods in the stores, money in the ATM machines, and videos on our Tik Tok for-you pages, all to keep us fat and happy. Sure, getting those things there was super-duper complicated, but that doesn't matter at all, because guess what? None of us humans ever have to worry about one bit of it! It all just takes care of itself. So, c'mon, what could possibly go wrong?

What indeed. Obviously, we can have circumstances beyond predicting where parts of these systems can simply fail. The more complicated something is, the more ways something can go wrong. That is the reasoning behind a very old piece of military wisdom that goes by the acronym of K.I.S.S. This stands for Keep It Simple, Stupid. Meaning don't over complicate the damn plan, make it as simple and straightforward as possible. The less moving parts and variable in an operation means there are less things that can go wrong. But our modern civilization has done away with this principle in favor of increased efficiency. At least, increased as long as things go as planned, and the world remains stable. We've had a few opportunities lately to see examples of what happens when things do go wrong, which they inevitably always will, and in unpredictable ways.

These last few years we have experienced the struggle of supply chains under the pressures of the pandemic, leading to shortages and misallocations of everything from PPE to toilet paper. We are still seeing those effects even as other shocks begin to hit different systems as the result of war breaking out.

What's worrying is that while neither of these have resulted in catastrophic failure, at least not yet, and the networks are making efforts at recovering, it will take many years of expert analysis and research to figure out even what went wrong in the first place, precisely because of how complex these systems are to understand. We have no idea how systems are going to react, if they really are going to recover, or what could happen in the event of another devastating shock sometime soon.

In addition to the prospect of failure, we also have to worry about these systems working too efficiently for our own collective good. These networks were built originally, and they evolved since, to be as efficient and effective as possible, and we have abdicated a lot of our decision-making authority to them in order to achieve that goal. It is not quite Skynet yet, but it is too damn close for comfort. But one thing we have not bestowed upon them is an ability to make ethical and moral decisions when making their calculations. And, for all intents and purposes, that is not a design flaw, it was the intended purpose of the system we created. We want the speed and the profit; we don't really care how it is achieved.

The global supply chains are a perfect example of this. All of the massive-scale industrial engineering projects exist primarily because global wealth inequality makes it cheaper to have stuff made in certain regions, even if that means you have to ship things halfway across the globe to sell it for a profit. Like having pears grown in one country purchased and then transported around the world to another country to be processed and packaged, all just to be shipped back to close to their origin to be purchased in stores by consumers. Often, what you get is a price for a can of pears to be considered negligible and cheap to the consumer, but for an amount of money that could mean an entire day's wages for the people who were doing the processing of them. By leveraging these major gaps in wages and standards of living as efficiently as it can, the network is actively enforcing that global inequality.

The same also goes for the global financial markets, which have a laser focus on creating wealth and growth, regardless how many workers might fall by the wayside, how many retirement plans might be destroyed, or even how much climate-wrecking greenhouse gases are emitted in the process. There is zero concern for any consequences not felt directly, and immediately, by the market itself. As long as revenues for the quarter are not jeopardized, then it is of no consideration.

Giving up our control of all of these processes to the hands of vast and unthinking networks not only risks those networks suffering catastrophic failures, it also threatens civilization itself. If we are unable to individually understand or influence anything more than very small parts of these systems, that means the same is also increasingly true for corporate leaders and governmental policy makers. Politicians and voters alike have less and less control over how any of these networks run. At best, they may find themselves merely managing very small parts of them, and often without the technical knowledge to really do that, and they certainly don't seem to be able to make significant changes to any of these networks, which are mainly owned by private corporate industries anyway, often ones that the politicians are beholden to already. Even though these systems have a very direct and significant impact on their national populations, there is simply nothing they can do. The system is literally too big to be affected by any small group of people, it has a momentum and inertia all its own. At best, even the most powerful among them are nothing more than data entry clerks for a complex, global system that nobody fully controls. There are hands on the wheel, sure, but that wheel is much to big to really be turned too far in any direction. All one can really do is give a tiny bit of correction over time, and hope that the next leader to come along doesn't reverse that direction again.

Collapse: The Rapid Reduction in Complexity

I believe that part of what we are seeing in the world is a result of us having reached "peak complexity," at least as far as our own species maturity level can handle it. We have been reinforcing such things as immediate gratification among ourselves for so long now, that we have a society where many people suffer toddler-like tantrum outbreaks whenever they do not get their way right now. Think about this the next time you encounter a "Karen" at the local Starbucks or grocery store and observe the sheer force of mental breakdown displayed if they dare to not have her cranberry bliss bar ready, on time and as ordered. As our world and the systems we have created to assist us becomes more complex, there is a rapid loss in our own mental ability to assist ourselves in the face of the absence of those systems.

The increasing complexity and hyper-connectivity of our globalized system of economics and industry has accelerated humanity's negative impact on the Earth's system and greatly reduced our resilience to ecological collapse. A small shock in one region is now able to spread easily and swiftly around the globe, growing in seriousness on the way, and having an unpredictable effect on other systems as it goes. This leads to a much higher risk of systemic and catastrophic collapse across all systems given the interdependent nature of everything. We have already seen these types of ripple effects, such as the shocks to the food system linked to the Arab Spring revolts, and the massive upheavals of the COVID-19 pandemic, followed by still more shocks we have not even fully felt yet from this new war. As a result, there is an increased risk of a worldwide "de-complexification" event, in which societal and economic complexity could undergo widespread and dramatic collapse.

We don't know much about how these shocks propagate through our economic and human systems, and so it is difficult to quantify our true level of resilience. What we do know more about is what would happen if we failed to create the required resilience in time.

The collapse event could be classified as a widespread reversal of the trends of recent civilization, involving not just the collapse of supply chains, international agreements and global financial structures but the complete breakdown of the rule of law and social order as well. This could be happening already, as part of a long, gradual descent playing out over many years as the result of many shocks to the system over time and the general decay therein. Or it could occur very rapidly, in the space of less than a year, due to a major "final straw" type of shock that creates a wave of disruptions to fast and hard to compensate for, resulting in cascading failure across the board. Only one thing is for sure: it is happening. Sooner or later, fast or slow, this ship is going down.

You would think something would be getting done to prevent such occurrences, and you are right, we are doing things. The problem is that the actions we are taking are the same as those regarding all the other factors like climate change, contagion, and conflict. We are still trying to keep what we have. We want to continue with our quality of life and "get back to normal." In short, instead of preparing the lifeboats we are locked into the struggle to keep the Titanic afloat, and we are unable to grasp the futility of such a strategy. It was a glorious ship, a magnificent bit of engineering and achievement, but it is going down, and whoever doesn't have an alternate floatation plan ready when it does will go down with it.

7

IT ALL COMES TOGETHER IN THE END

Convergence

There is actually one more "C" we need to discuss here, and that is Convergence. I know, I left that out at the beginning, but that was to help you come to this conclusion on your own.

What we are seeing in the world right now, from the "faster than expected" reports about the worsening effects of climate change, to the increasing spread of war and looming famine, and even to the accelerating breakdowns of social bonds and individual mental health, all of it is coming together right now. Everything I have written about, and so many things that I left out, it is all working upon the world together at this point. All of these disasters and changes are converging upon one another at the same time.

That idea of convergence, of the perfect storm, is what makes a rapid and total collapse of global civilization not just likely, but inevitable. We can no longer address these problems separately, which would have once been within the realm of human endeavor. Now, any action we take to fight one or even two of these looming crises will simply exacerbate the others to the point where they will consume us from a different direction. How can we address emissions when at the same time we have to fight a war? How can we make a transition to renewable energy when at the same time the economy is hanging by a thread that needs to be reinforced? How can we stop deforestation when the farmland we have is already not enough to provide food for the people?

It goes on and on, these problems have been ignored for so long it is now too late to address them singularly. It is like a cancer. If you detect it early, and then take steps to address it, there is a very good chance that it can be beaten. But if you leave it to grow and spread…eventually it is everywhere within the body, and there is very little chance of surviving it.

COVID was a real kicker, for sure. That little bug came along and exposed a bunch of festering, but not yet noticed, problems in our world, from medical unreadiness and supply chain vulnerabilities all the way down to our own individual distrust of one another. It is my opinion that the coronavirus pandemic can be seen as the trigger event for the sort of collapse I describe in conclusion.

But it didn't stop with COVID, oh no. While still being strangled by the pandemic issues, we went and kicked off a new major war in Europe, right smack in the middle one of the world's largest breadbaskets. And, funny thing, it happened right at the same time as the IPCC was delivering its most stark and dire warning yet about the looming specter of the climate crisis bearing down on us, which we are all starting to feel already.

Climate, contagion, conflict, and complexity are all showing their ugly sides to the world. Having read this far I am sure you have started to see by now how all of these things seem to not just be happening at the same general time, but also all seem to play into each other. Parts of one accelerate parts of another, and slowly the interaction of them all has created a perfect storm. Resource scarcity leads nations into conflict with each other. Loss of fish in the sea exacerbates the food shortages. Need for farmland accelerates deforestation. All of the things we have talked about, they all affect each other, and for the worse.

As I write this now, we are facing so many troubles that it is hard to keep up with them all. Whoever said bad news comes in threes was wrong. It comes in dozens. Economic instability, active and soon-to-be active conflicts, supply chain chaos, environmental disasters, emerging world famine, resource shortages, massive energy crunch, wide social divisions and cultural disputes, a deadly pandemic, geopolitical reshuffling, wage inequality, civil uprisings… The list is a very long one. In fact, if you can name it as a problem, it is probably happening right now to some extent. We are not just experiencing one or two, we are getting hit with everything at once.

This isn't just the Four Horsemen; this is a full-on cavalry charge.

There is an incredible number of factors to consider when it comes to collapse. Any one of these things, if it gets to the extreme, can be the thing that takes us down. And we are facing them all at roughly the same time. What is almost impossible is to try and see how these occurrences can also intensify and feed off of each other. How will they interact? Our society has grown incredibly complex and interconnected, and everything depends on everything else. At the same time, and for the same reasons, everything can also affect everything else.

The issue is that most of the information out there seems to be divided into multiple camps and varying visions of how the world will face collapse and what it will look like. Slow and gradual, driven by a worsening climate crisis? Fast and violent, like a nuclear war? I will spare you the listings, we have gone over them already, but rarely do I find much information about how all the different possibilities could intersect and interact with one another. One of the challenges is squaring up the knowledge of both the long- and short-term effects of the climate crisis with the ever-present demands of daily life. Another is defining how effects from one event can ripple across the landscape of society and cause other events to unfold. Or, how humanity will react to the various changes that will happen in their lives as a result.

We all know, to some degree or another, how disastrous the effects of climate change will be for the natural world. But what most of the scientific models fail to take into account is the correlation those changes to the natural world will have that is reflected in the artificial world. Few indeed will confront what climate change means for our social, economic, and political systems, and for human interaction within those bounds. And, as of yet, I have seen none that go into detail about the irrational behaviors that many people tend to exhibit in reaction to any given crisis, or how we often respond to events not with practical plans for resolution, but with ones that are politically expedient or economically sound. At least, they seem so at the time. It will be humans that push the button to begin a nuclear apocalypse, not mother nature. She may provide the impetus, but we will be the ones who make the moves.

In many circles confronting what climate change means for our economic, social, and political systems has taken on an increased sense of shared urgency. Recent studies have shown that an increasing number of people view our climate-challenged future as so dark that it has become a major factor in their decision not to have children. They worry about the effects of climate change on their health, both mental and physical. They believe that climate change will hurt them personally in their lifetimes and, at the same time, they have little confidence that political efforts or technological solutions to prevent climate change will succeed, or even help at all.

After many decades of leaders from around the world coming together to try and work out solutions for dealing with climate change, emissions are still rising, the ice is still melting, forests are still thinning out, and biodiversity is still disappearing. And still, we go to war. How's that for inspiring confidence?

In addition, while policymakers, climate scientists, and environmental groups have been engaged in a half-century or more of conversation about the future of our world, most people on the planet still do not see any emergency looming. The science of climate change has failed to a spectacular degree to emotionally connect with most of humanity, rendering leaders ineffective despite all the warnings of what is coming. On top that, most national leaders are burdened with the need to bow to the desires of the lobbyists of their corporate and wealthy sponsors, in addition to fielding the political requirements to maintain party dominance and work for re-election. Most of their efforts are directed towards such by necessity in the political systems we have, and unfortunately there are no real sponsors of anything but "Business As Usual."

That warning we have been ignoring is basically that if we do not curtail the emissions of greenhouse gases into the atmosphere, it threatens the viability of modern civilization, results in catastrophic degrowth due to overshoot, and just worsens a mass extinction event that is already in motion.

Alas, this warning - as stark as it sounds – is just not really connected to the complex human systems of society, such as finance, telecommunications, and transportation, not to mention the very basics of food and water supply. These systems are just left to continue to evolve and operate as if the threat of climate change does not even exist. Humans are wired to make the short-term a priority, and so concepts such as "tipping points" are left on their own as distant and abstract to the technicalities of daily life.

This failure to draw the right picture for everyone means that humanity has now entered into an uncertain future without even really being aware of it and continues to spew out emissions many times faster than ever before.

We all need to continue to improve our understanding of the science behind this potential catastrophe, sure, but what we really need now is to start looking at how our artificial systems of civilization are intertwined with the fate of the climate of this planet. The real story about our looming emergency must highlight how vulnerable society is to near-term climate disasters, and how we as people are going to react to the shocks of events as they unfold. In addition to the climate ones, there are also human "tipping points" and "feedback loops" which reflect people's actions to the ever-increasing pressures of a world that is reaching its limits.

People react to the changes in the world around them. Often unconsciously, especially when they cannot grasp the meaning or causality behind those changes. For someone who is either ignorant or in denial of climate change, how do they perceive the resulting effects reflected by increasing disasters, rising temperatures, and frequent bizarre weather events? How will they react, and what effect does it have on their mental health? Socially, culturally, and religiously, everyone sees things in different ways, and while we are all having the same experiences, they are given different contexts in different minds. We may all look at the same effect in the world, but when asked we will all have many different beliefs as to what the real cause was.

One thing is for sure is the mental, physical, and emotional stresses these changes place on everyone. Such stress is aggravated for those who do not understand what is happening, and thus they tend to react in unpredictable ways. It begins to reflect in their everyday life, in the decisions they make, and the attitudes they exhibit to others around them. They can become irritable, unhappy, and have resentments towards the world in general that they may not even recognize or understand. Those feelings build, and can cause them to act irrationally, or even lash out at the world and the people around them.

Have you not noticed the general "weirdness" of people lately? That Mass Societal Discontent we talked about? Have you seen how people are becoming more confrontational, quick to anger, and impatient with even the slightest inconveniences? Look at the statistics for the increase in occurrences of things like domestic violence, incidents of road rage, public freak-outs over shortages at restaurants, and even more serious violence like mass shootings. As things continue to progress, and our world becomes harder and harder to live in, how long before things reach a tipping point? How much more will it take before people begin to snap from the pressure in larger numbers? And when they do begin to snap, just how fast could that turn into the kind of mass hysteria that is usually only seen at the beginning of a zombie apocalypse movie?

Everyone has their own tipping point, I am sure. But collectively they all reinforce each other. That is the nature of a feedback loop. Each incident compounds the confusion and stress, causing more people to experience mental breaks, which in turn continue the process. Basically, the worse things get, the faster they get worse.

Leaders are not immune, either. Look at some of the drastic actions taken by politicians across the world lately. Look at the reactions of others to those actions. We have a full-on war going in Europe. Rising tensions between China and the US. North Korea is testing missiles at an alarming rate, the Philippines just elected the son of a notorious dictator, and in America we are busy stirring up new riots over abortion rights and seriously polling the population about the likelihood of a new civil war. The references could continue into an incredibly long list of inflammatory actions being taken across the globe by our so-called leaders.

Our latest war in the world kicked off literally in the shadow of the greatest warning about the climate crisis that the IPCC has ever put forth. We were told that the only way to stave off the worst ravages of climate breakdown can only come through a "now or never" dash to a low-carbon economy and society. But it didn't take us but a few days to toss that warning right in the trash and double down on fossil fuel use. Nations are even bringing back coal power of all things, and they are chomping at the bit to do so. They are not doing it for nefarious reasons either, but because they literally have no choice. Demand for electricity is rising as supply is decreasing, and the situation is only going to get worse. Collapse now or collapse later? What kind of choice is that?

Energy crisis. Increasing climate disasters. Ongoing war and looming conflict. Social unrest. Supply chain breakdowns. Global food shortages. Emerging diseases. Widespread droughts. Pollution. Inflation. Rising homelessness. Plastics in everything, even unborn children. The list goes on and on. Systems are breaking down left and right across the world and the incredibly complex systems we have created for ourselves. How much more straw can the camel take?

Cascading failure.

That right there is the monster lurking in the dark. Tipping points in our planet's system – such as deforestation, the loss of biodiversity, and the melting of ice sheets – are certainly existential long-term threats. But they are already causing increasingly severe weather events that will soon become more extreme and frequent enough to trigger what is known as a "cascading failure" within our structure of global society. Added to this are all the things mentioned above. We are not facing a single crisis. We are being bombarded by crises from every direction and angle, all of them feeding off each other. Systems are beginning to buckle. That is what so many people are missing about everything.

Cascading failure is when multiple shocks across both manmade and planetary systems lead to catastrophic disruptions in their functioning. These disruptions, given how much our global system is interconnected, can affect a single nation directly but also lead to the failure of global supply chains and financial systems across the world. The entire global economy could come apart in the blink of an eye. The COVID-19 pandemic showed us a preview of this in action. And it is hardly limited to economics. Everything is tied into everything else, and it is a recipe for catastrophe. Some things can be predicted, and even planned for, but things like war are incredibly unpredictable. Energy, food, finances, everything is affected. As the planets tipping points are passed, so too are our human tipping points rapidly approaching that mark.

Peak complexity is real. Our modern world is a very complex and interconnected one. It may not seem as such while we are busy going about our daily lives, but ours is a structure of interdependent systems, in which the failure of one part can easily trigger others, and those, in turn, affect more, until the entire thing collapses in a massive cascading failure event. Cascading failures occur when one part of a system fails, leaving other related parts to compensate for the failed component, or to "take up the slack." This then brings these compensating pieces closer to overload and failure themselves, potentially causing them to break down as well, prompting a string of additional failures down the line, one after another.

Civilization functions as a diverse infrastructure of systems, all coupled together and dependent on each other for functioning. Due to this coupling, interdependent networks are extremely sensitive to random failures, and even more so to targeted attacks. The failure of a small number of parts in one system can trigger a cascade of failures throughout the entire network and bring it all to a screeching halt.

In a healthy and well-managed system, the complex interdependence provides a benefit that allows for all the conveniences of our daily lifestyle to function. The problem is that this complex system of "just-in-time" logistics chains, and everything being maintained for "instant access" and "on-demand" has been designed around the assumptions of a stable world in which everything moves smoothly from actions to predictable outcomes. But it has been built on top of a hugely unstable and complex platform, which is our increasingly climate-disrupted planet.

So, what happens if the system begins to become unhealthy, unbalanced from the effects of this unstable platform? If it is also perhaps mismanaged, or stressed by outside forces? It can still function, for a time, but is slowly begins overloading, and when operating at the capacity redline, well, it is only going to take a single significant event to start the cascading failure across the board. That is where I believe we are, right now. Our systems are overloading, they have taken everything they can, and the entire thing is beginning to become catastrophically unbalanced.

We have already seen the beginnings of this. Our system was humming along just fine, at least so it seemed and for the most part. Much of the mismanagement was kept from being apparent to the general public, and the real climate effects are just beginning to have an impact. But then along came COVID-19 and the pandemic. Very quickly, this event began to expose the weak points in our systems and has slowly ramped us up to that redline. Parts of it are now very close to collapsing individually, and other parts are already taking up the slack for the failing ones. There is precious little reserve capacity left. Then we had the invasion of Ukraine by Russia, which has grown into a war which could drag on for years, maybe even spark the 3rd World War, or result in the use of nuclear weapons as things deteriorate.

As it sits now, it would only require a single trigger event of a sufficient magnitude to cause the whole thing to come apart around us.

What could that trigger event be? Well, due to the state of the world and our current global instability, there is a myriad of possibilities. War between the United States and China, or NATO vs Russia? A terrorist attack with a bioweapon? The "Big One" finally hitting the U.S. Westcoast? A new viral pandemic or the emergence of bacterial resistance to antibiotics? The long overdue repeat of the Carrington Event? Imagine for a moment, with everything going on in the world right now, what would happen if another "unforeseen" event kicked off in the next year?

It is a long list. And it grows longer every day, it seems. And the fact is, that trigger event may already have happened. The pandemic may have been it, and we are just watching the world slowly come apart around us.

There are a great many studies about the looming specter of climate collapse. And most of them, while quite accurate, were done pre-covid. None of them really considered the possibility of a major war. Few, if any of them, really take much more than a passing swipe at factoring in the "human nature" element. As all of these horrible climate-related disasters begin to take effect and pile up around us, how are people going to react? Individuals, political leaders, whole segments of the population. How are they going to handle the changes? Will they just sit back and let those long-term scenarios play out according to the scientific models? You really only have to look at how we have all handled the pandemic to see that the general answer to how we will handle it all is "not well."

In the near future, the food crisis is likely to get much worse. The risk of multi-breadbasket failure is becoming significant, and it only goes up as we hit the threshold of global warming above 1.5 degrees Celsius – something that could happen as early as 2030 should emissions continue unabated. Skyrocketing food prices, leading to civil unrest, mass starvation, and death. And here we are, engaging in a major conflict right in the middle of one of those breadbaskets. How's that for an accelerant?

One of the most worrying things about all of this is that compared to the long-term climate studies and models, we don't really know a whole lot about just how fragile the various parts of our clockwork global systems are in the short term. We just don't understand as much as we should about how our manmade systems will respond to these events, and other occurrences, which will only be happening with greater frequency as the climate destabilizes further. How are these things going to trickle down the line, and what about the effects on the most vulnerable and politically unstable nations? Finally, given the general tendencies of human nature, how are the people affected going to react as things get worse? What kind of human feedback loops might emerge that can cause totally unforeseen events of their own, things not predicted by any models due to their sheer irrationality?

These are not really the distant existential threats raised by our current and abstract models of the ongoing climate crisis. They are urgent questions that people have avoided for a long time, and they need answers. Unfortunately, I believe those answers are unpalatable to everyone. Mostly because the answer is that the collapse of global civilization is inevitable. And I believe that societal collapse will occur in advance of the ecological collapse, mostly due to our reactions to those increasing pressures and the failures of our complex human systems. The Earth is a pretty resilient machine, but our creations are quite fragile.

And so. It is this convergence between climate change effects and our human reaction to them that is my focus. The issues of droughts, famine, migratory climate refugees, extreme weather disasters, and the rise of open warfare. All of these and more are things I hope to spread information and awareness about with this book, as well as my blog of the same name. It will be a journey of learning for all of us since I myself still have so much more to understand. This book details what is happening in our world, but it is not about getting into the nitty gritty of post-collapse realities, or even hopes for future rebuilding. Those are subjects for future books, assuming I will have the time to write them, which is certainly not guaranteed. Here, I simply hope to help bring people to the understanding and acceptance of what is about to happen. It really will be the end of the world as we know it, and before anyone can even begin to know what that will mean for them, they have to know what it really is.

The primary takeaway of this book, and the independent research I hope it spawns in you, dear reader, is that societal collapse is inevitable and that it will happen sooner rather than later. It will not be the slow, gradual decline of the Roman Empire, but a more rapid and violent descent into chaos the likes of which the world has not yet seen. Never before has a society existed that was as complex and interdependent as ours is now, and its fall will be a unique event to this planet. As for all the scientific models, I don't believe they are entirely predictive. Strictly from a scientific standpoint, sure. They all do quite a good job at detailing what awaits us in a climate-challenged world. But the models do not account for the reality that people will probably react irrationally to escalating crises and mounting disasters as they begin to pile up. And many of the processes of advanced civilization will simply not allow themselves to be interfered with, as has been demonstrated by the decisions of political and business leaders for quite some time now. Business-as-usual guarantees the end of the world as we know it: our current way of life is not sustainable. It never was.

I am sure, once you begin your own journey into all the threats the world faces, you will learn about all the things that I left out of this book. There is more. Much more. What I have detailed here is only the tip of the iceberg that we are about to ram into at full steam.

A rapid collapse of civilization has never been the most likely scenario, but I would argue that it has never been more likely than it is today, and the chances of it happening are increasing daily. The reality of the situation is that not everyone will survive. Not everyone can survive. But those who are ready for it will have a better chance than those who neglect to prepare at all, and part of preparing is knowing exactly what you are preparing for. In this case, that is the total collapse of civilization.

The Final Word

This new Anthropocene is certainly the Age of Consequence, where we all get to experience the drawbacks to all the benefits enjoyed by the previous generations let stuck us with the tab for their wild party. We are all now living in a time that I think history will remember as the beginning of the fall of civilization. And unfortunately, I think that history will be written by hand. What do I mean by that? Just take a look around, mate. The world might still be spinning, but things are really starting to come apart fast now.

From devastating heatwaves to a new World War, a looming economic apocalypse to the beginning of a worldwide famine. We are really in the thick of it right now. "Faster Than Expected" is a term that has become a running joke in many circles, because of the sheer number of new studies emerging showing that we are already experiencing catastrophes that were originally thought to not be on the table until 2050 or so. Everything has sped up, and the hits just keep on coming. We have a couple dozen catastrophic threats bearing down on us all at once, and each of them is enhancing and accelerating the others. Look at Europe right now, the devastating heatwave causing fires and heat deaths all over, especially in Portugal and the U.K. That type of weather was expected for 2050, not 2022.

Climate change used to be a more abstract idea, something that would affect us far in the future. Because of that, no one cared too much, not really. Protests and activism were very fringe movements, and the science was still a bit spotty in the early days, a fact we are starting to realize now as things begin to occur a bit sooner than we thought they would. Perhaps the tipping points have already happened, maybe we missed something in our calculations and the feedback loops have already started an irreversible process. That's the problem with science, of course. We just don't know everything yet. Every day of every summer, you can literally feel the planet getting hotter. Feel it on your skin. In your sweat. You can see it in the strange, surreal, flaming sunsets.

But now, climate change is finally here, live and in color. We can all experience the weight of it just by stepping outside. Every day you can physically feel the world getting hotter, see the greens turning to brown, and taste the pollutants in the air. The consequences are here, today. Our planet is being roasted alive, and we are living at the very raggedy end of civilization. A lot of people say it, "man, it feels like the world is ending!" Well, guess what? The reason it feels that way is because that is exactly what is happening. These things you are experiencing in the heatwaves and wildfires, in the food shortages and the floods, these are not anomalies. They are not warnings or random disasters or temporary states we have to deal with before we get back to "normal.". Not anymore. This is the way things are now, and only going to get worse from here on out.

Normal? You want to get back to normal? Well, sorry, but you can't get normal anymore. Normal is out of inventory. The stores no longer stock it. There is no more normal, and the best you can get is what you have right now, for as long as that may last.

Many of the things we are going through right now used to be thought of as problems for someone else, not us. Third-world problems. But now, in the so-called "first-world," we are starting to feel it. Energy grids are stressed and failing. Inflation is running rampant, and we are seeing shortages of everything from food to car parts. There is a war on in Europe, and in America the government of New York just put out a public service announcement about how to survive in the event of a nuclear attack. In many parts of Western Europe, the leaders are telling the people to start getting ready to chop wood to heat their homes this winter because there may not be any power. Every day there are new catastrophic floods, fires, tornadoes, and other "once in a century" disasters, but they are happening regularly now. Utilities are already practically begging people to cut power consumption. Us modern "developed" westerners don't understand that in the poorer and less developed parts of the world, the future is now, and has been for a while. People are experiencing deadly heat all the time, it is really killing people, and they have no way to stop it. Parts of the world have already collapsed into chaos, and just because the news doesn't like to show such things to us doesn't mean they are not happening. The media and government don't really want us to see what is in store for us all in the next few years, because we might panic. And well we should.

The future of climate collapse is already here. Today. Nations are already going to war as the resources dwindle. People around the world are already rising up to tear down the governments that no longer provide for their welfare. And most of us in the more stable parts of the world don't really understand that as our civilization continues to devolve and crumble beneath our feet, we are all on the way to becoming refugees with no place to flee to. Think about it. Go ahead and imagine living through the kind of accelerating disasters and spreading conflict that you usually only get to see in movies or read about in fiction. The power grid is gone. There's no clean water. The harvests withered long ago. You are completely without food or shelter from the devastating heat. People are fighting everywhere, some at the direction of what warring powers remain, other just to survive and eat for one more day.

That's the future. And it is not a far off one.

As I write this, in preparation to publish within just a few days, even the mainstream media is unable to hide the carnage from us any longer. We have recently seen record-breaking heat waves across the world, many spawning wildfires that ripped country sides and towns alike. The U.K. reached 40.3° Celsius, or 104.5° Fahrenheit, for the very first time ever, but quite obviously not the last time. China is enduring a blistering heat wave as well, with the government having just issued its "highest possible" warning for excessive heat. Thousands of people across the world have died from the heat, just in this last bit of July 2022. Millions more have been displaced. Ongoing droughts have worsened greatly, destroying crops and killing livestock, creating the potential for a disastrous famine, and this time it won't just be in Africa where people have traditionally ignored it. This time it is going to hit home.

Even in America right now, over 90 million people are under heat alerts in more than two dozen states. Violent thunderstorms and tornadoes slash across the landscape, floods are wiping away entire towns, and almost every day brings a new disaster declaration somewhere. In the southwest, Lake mead has seen water levels dropping rapidly, and that spells disaster for many millions of people across California, Arizona and more. Wildfires have laid waste to hundreds of thousands of acres across the nation, and the season is just getting started, and all of it is amplified by extreme or severe drought that shows no signs of letting up. In Texas the power grid is overheated and overextended, with many people already suffering from rolling blackouts as the utility company tries to keep the AC powered. California is telling its residents to try and limit charging of electric vehicles during peak hours, and this among record high gas prices. The grid is threatened due to heat waves, but apparently they did not foresee this when they wanted everyone to convert to EVs? Running steadily at max capacity, with maintenance put on hold...just how long can the aging infrastructure take the pressure? No one seems to see that it is not just the source of energy that is the problem, the problem is that the *demand* for energy is too high.

The war raging between Russia and Ukraine, along with the fears of its spread and the possibility of new wars involving China or Iran have added to the carnage. Energy scarcity has driven costs through the roof, and soon much of the EU nations, especially Germany, are going to face a cold winter and the death of their industry when the gas is cut off from Russia. The economic damages of the pandemic have joined with those of climate change and conflict to create monumental upheaval.

Could it really be that we did not see this all coming? Were we not warned over the last 50 years or more? Why didn't someone tell us about the consequences of our actions long ago?

The truth is that we have always been able to clearly see the steady environmental destruction doled out by our industrial activity. We have watched ourselves drive species after species to extinction, we observed the steady decline in the Amazon rainforest and the oceans of the world, and we even recorded it all with great precision. And yet, we just kept right on doing it. Systematically grinding up our world and turning it into dollars and cents in ledgers at banks all over the planet.

Those consequences will be wave after wave of mass deaths across the globe. Millions upon millions of refugees from climate disasters and conflict alike. Crashing economies, shattered societies, resurgent diseases and new novel pandemics. Crushing famine and drought, soaring costs just to survive for those few who still have the means, and a dystopian hellscape for the masses. It will truly be a wasteland. We have fucked around for quite some time, and now we really are about to find out.

We are starting to realize that reality a bit now. Like I said, even the mainstream media has finally started to tell the stories. The heat is killing us, not in a hundred years but right now. The desperate response so far across the world is to burn more coal, to reopen those few plants that were closed, and extend the life of nuclear power plants to generate even more power to try and stay cool keeping both ourselves and the economy alive. This is quite obviously a knee-jerk and self-defeating reaction. The problem, as you would think we would recognize, is not that we don't have enough energy, it is that we need too much energy. The demand for it is too high. We are caught in the ever-increasing loop of a heroin addict, always looking to get that high, and always needing more than last time to achieve it, and still more after that. Rather than fight the addiction to fossil fuels, we are feeding it harder. I don't know if you, dear reader, have ever witnessed or experienced extreme heroin withdrawal, but it is a horrendous thing. Once the energy shortages and power outages really begin, the killing heat will exact a grim toll on humanity and civilization.

"No nation is immune. Yet we continue to feed our fossil fuel addiction," UN Secretary General António Guterres recently stated when addressing the crisis. "We have a choice. Collective action or collective suicide."

It would seem to me that the choice has been made. Suicide it is.

The Anthropocene is accelerating, and the rate is much "faster than expected." The world's forests are being burned and cut down much quicker than they can be replenished. Wars are spawning and tensions are escalating in ways no one could have predicted just a short time ago. Sea levels are rising three times faster than predicted. Political division between parties is widening and social cohesion is breaking down. The permafrost is melting, and the icecaps are vanishing at rates that were unforeseen by even our best scientific models. Economies are crumbling, resources are growing more and more scarce, and entire species of plants and animals are dying out in droves. Collapse is not a linear thing, it is exponential. And it is here now.

Even if we stopped carbon emissions today, every last bit of them, carbon dioxide concentrations will continue to climb past the current 420 parts per million to about 550 ppm, simply because of the heat trapped in the oceans, and the past emissions that have not even caught up with us yet. Global temperatures will continue rise, no matter what, right into the catastrophic levels. Resources will grow even more scarce even if we only use what is directly needed for survival. Wars of the most horrible kinds imaginable will be fought for those dwindling commodities. And while we fight, the earth will continue becoming more inhospitable to life itself.

And that assumes we actually did something drastic to confront this crisis, that we actually took action that could stop all emissions. And it would have to be that dramatic, because right this minute we are already approaching the tipping point of 2 degrees Celsius when the ecology of the planet will become so degraded that nothing will be able to save us.

We are certainly not going to take such action. Not even close. At this point, it is almost entirely impossible to do so. Those with power and control over things have always denied the reality of the climate crisis and did nothing to stop it. Us regular folks were led directly into this catastrophe, like cattle. Extensive heat waves and record droughts. Escalating wars and rampant pandemics. Chaotic weather patterns and shrinking crop yields. Melting ice caps and disappearing glaciers. Raging wildfires and crushing floods. Mass migrations and expanding desertification. Political turmoil and civil unrest. The extinguishing of sea life and rising ocean acidification. Spreading famine and rising income inequality. It goes on and on. The feedback loops have already started. Already we are seeing the occurrence of one catastrophe worsen the effects of some other catastrophe. The breakdown will be a cascading one, nonlinear and exponential. These things happening in our world right now, they are not just isolated disasters. They are interconnected precursors. Harbingers of the common future that awaits us all.

The urban sprawls of cities will become deathtraps. Social cohesion and the rule of law will disintegrate as community breaks down and the governments have less resources, and less individual will, to fight the process. It will get worse and worse as things progress. This is already happening in many parts of the world, and you may have even seen it in the so-called "first-world" where you are right now. Rising violence and crime, the increased mental and emotional instability among people, and the formation of cult-like groups and tribes with ever more extreme and bizarre agendas. Whatever is left of governments after they are done warring with each other will institute ruthless security measures, backed and enforced by heavily militarized police forces more thuggish than they are now. Developed nations will turn into isolated fortresses, cracking down on all freedoms and rights in the attempt to ward off climate refugees and prevent the uprising of their increasingly desperate public. The ruling class and wealthy elites will try and consolidate their power into smaller areas at first, perhaps something like city-states, and later maybe even arcologies within those areas, Judge Dredd megacity style. Finally, as even that begins to unravel, they will retreat to isolated and defended compounds where they will have all the supplies needed to ensure their comfort for what remains of their lives. But the rest of us will be fighting over the scraps in the ruins hunting for food, water, and medical care. The dystopian wasteland, realized at last.

But not for long. Even that iron-fisted type of order will give way soon enough. If it even gets that far without us having annihilated the planetary population via nuclear exchange, which I see as more likely than the authoritarian dystopia many envision. I do not think nations and powers will simply submit to destruction and defeat without first having used every weapon at their disposal. There is a precise mathematically derived number which shows how many times such a defeat and absorption has happened in history. That number is "zero." Nations that have them will most certainly use their nuclear weapons as a last-ditch attempt to stave off destruction, or at the very least to take their killer with them into that dark night.

There are many different scenarios that have been scientifically mapped out for the future, but once you come to the conclusion that nothing will be done to stop "business as usual" there are only a couple that really dominate the probabilities. In one, there is a massive die-off of something like 80% or 90% of the human population, which is then followed by a long and gradual ecological stabilization. This is the wasteland scenario. In the other, it starts like the first but doesn't stop at 90% population loss, it continues on with the speedy extinction of humans and most other species on the planet.

There is another, slightly better scenario than the wasteland one, but it is dependent on embracing degrowth, with all the mass population reductions such entails. A vast ramping down of all economic activity, a halt to the production and use of fossil fuels, switching over the world population to a subsistence style diet with an end to industrial agriculture and a return to local farming, and finally, complete cooperation between nations to achieve such things peacefully and immediately.

Yeah, I am sure you see why that last one isn't very likely. And besides, the future there is pretty close to that dystopian life of drudgery we described earlier, because remember: we already have quite a bit of global heating "baked in" to the equation no matter what we do.

We've known for many decades now what harnessing fossil fuels and basing civilization entirely around them would do to the world. But the potential for profit and growth from fossil fuels, and the lifestyle the burning of fossil fuels afforded to us all, overrode any chance of a rational response to that knowledge. We also knew that by the time the looming effects were observable and measurable by us, it would already be too late to really do anything about it. We are there now, and runaway catastrophic climate change is certain to speed up dramatically.

There's not much that is realistically doable to stop collapse or even mitigate it at this point. What grand ideas we do have are fantasies because they are based around sweeping changes and an incredible level of cooperation and sacrifice, not just by people but by nations as well. Dramatic changes to entire concepts of economic, political, and cultural norms that have become so ingrained in our collective being that most people cannot even grasp them conceptually, much less actually put into practice. And now, as we enter into a new multipolar world of national independence and constant conflict, there is just no way that any country, especially one of the primary powers, will be willing to look beyond its own salvation and leave itself open to aggression from the others. For individual people and groups, the divide on various issues is already so deep that any occurrence in the world becomes a blame game and a way to attack each other. Whenever something bad happens, immediately the war of words begins, with the right and left, black and white, religious and atheist, red and blue and green all blaming each other and jockeying for position, and meanwhile nothing really gets done about the actual problem.

What is much more likely is that the continued pressures coming to bear on the planet and on civilization will spawn more and deadlier spinoff conflagrations. Small disputes between people are increasingly going to turn into major ones, causing people to snap under the strain. Citizens are going to react with rage at government's inability to deal with the numerous and unpredictable crises that are emerging. Believers of various types will stir up religious fervor, and we will certainly see a dramatic rise in doomsday-type cults. There will be an increasing level of hatred and violence towards minority groups and migrating climate refugees. Escalating conflicts over resources, particularly food, energy and water, will grow to be more common, at the neighborhood level as well as on the national stage. At the same time, we will see the good parts of people become less and less apparent. Decreases in empathy and generosity, kindness and goodwill. We are going to become more tribal, in ever tighter and smaller groups, and more vicious in defense of our own particular narratives and conspiracy theories.

In short, humanity, and civilization, is going to enter something like a catabolic state, and begin consuming itself, accelerating its own collapse.

Civilizations have collapsed many times before all throughout history. Humanity has been building and living in civilized fashion for thousands of years, and our history has seen many large and prosperous civilizations collapse entirely. You could say it's really nothing new for us, and that is true. But this time is different, for a variety of reasons.

In the past, whenever a civilization collapsed, there were other civilizations around, and these would eventually rise to replace the fallen. While a nation may have depleted the resources around it through over exploitation, there was always some new and fresh place to go and conquer, to colonize and begin anew the process of exploiting resources once again. Furthermore, civilizations and nations were by no measure as connected to each other as they are today. While trade was a large part of things, there was very little interdependence. There was always "somewhere else", some new world waiting to be conquered and plundered, full of resources which would fuel the rise of new civilizations. And that is exactly what we did. We rampaged across the entirety of the globe, consumed everything in our path, and eventually we globalized our civilizations together into one big conglomeration. The drive for continuous and unlimited expansion and exploitation, spurred on to an incredibly accelerated pace by the Industrial Revolution, made us bend the whole planet to our collective will. We dug up and chopped down everything that could be seen to have the slightest value, polluting and destroying as we went, and we did so everywhere. We have made the world one big, connected civilization. True, there are warring factions within it, but by no means are there any truly isolated and independent civilizations anymore.

Because of that, civilization itself has become a death sentence. This time, there is no "somewhere else" we can run to. No more undiscovered country that we can plunder to fulfill our needs. No plot of pristine and resource-rich territory that can be our salvation. This time the entire planet will go down the drain. This will be the final collapse, with no new lands or people left to exploit, no new civilizations to replace ours. This is not a cup half-full or half-empty situation at all, it is completely drained. We have used up the Earth's resources, and soon we will be struggling to scrape by on a desolate world. We have "civilized" the entire planet, and there is no "Planet B."

The collapse of civilization destroys everything. Everything. The fundamental systems of our civilization depend on stability in order to even exist and function. They are already becoming unstable now. We can all see it, though many refuse to acknowledge it. Where I live in the Southwestern United States, there are already warnings that the water will begin running out, like really running out, as soon as 2025. That's not the future. That's a little over two years away. I visit the reservoir of Lake Mead all the time, I've been doing so regularly for almost two decades now, and I can tell you that it is going dry much faster than the government is telling people, and when it does, it's game over out here. Lake Mead supplies both water and power, through Hoover Dam, for the City of Las Vegas and others. As the water runs out, the lights go out too. And more importantly, the air conditioning stops. And then, you die. Good luck living in the desert with no water or power, especially after living with the delusion that it was all supposed to be palm trees, pools, glittering lights and golf courses forever. That illusion is going to shatter like brittle glass for a lot of people.

It is the same for you, wherever you are. There is something critical at risk that your entire area depends on in order to survive. Some resource or industry or chain of systems that is having trouble now. Whatever it is, you know that if it collapses, so does your city or town. this illustrates the problem of system failure within a massive network of interconnected and interdependent systems. This is how societal collapse happens. The dominoes start to fall. One system goes down affecting another, and then they both go down, impacting a third and fourth…you know how dominoes fall. That is how we will fall too. Cascading failure.

That cascade might seem like it's in some distant future, but it's not. It is beginning to happen right now, slowly, and will be at full speed this decade.

The key word in that sentence slowly. Yeah, everything we are getting hit with now, this is the slow part. The easy part. When the real collapse begins, it will be a sudden and rapid de-complexification of our entire world. Maybe the systems will just fall apart from the stress of trying to compensate for the continual damages. Law and order fall away from society like the masks they always were, and chaos reigns supreme. Maybe one nation or other finally faces its existential moment in the war and buttons get pushed sending much of the world right to the ash heap. Maybe a new pandemic emerges with a mortality rate closer to Ebola than the Flu, and we all get to sit around hoping to be one of the 10% who get to survive the aftermath.

You want to know where we really are as a civilization? That's where we are.

It's time for us to realize this. Not just individually, or even as separate nations, but as a species. The people running the show have led us right to the point we are at now. The climate is starting to bake us. We are going to war and starting to kill each other. Mutated viruses are spreading unchecked throughout the population. Microplastics and other pollutants are being found in unborn babies. We are getting ready to burn more of what little remains of our forests because otherwise we will freeze in the winter for lack of power.

And this is the early part. The easy part. Thanks a bunch, leaders, and thank you to the asshats who mined and drilled and burned the land to rubbish. And thanks, corporate billionaires! I love my new Xbox that I didn't need and couldn't afford! Now we're all going to get to play Fallout, but it will be the IRL Edition.

Yay, right?

We are the ones who have let our future be defined this way. It's easy enough to understand how we ended up where we are now. Just look at how climate change was viewed as this slow and manageable process that we didn't have to worry about until way in the future. Something we could deal with later rather than something that would kill our world and take our civilization with it. It happened like this because our leaders were easily corrupted, convinced to believe this lie by the economic interests that got them into their positions for the purpose of kicking the can down the road forever. And we all kind of fell right into line along with them. Because we liked the rewards so much. All of our fancy lifestyles and incredible conveniences, the vacations to the coast and the perks of having whatever we wanted just in time whenever we wanted it.

That's how we ended up here. And even now, that's how we all pretend it can be fixed. How we can pretend "net zero" or "carbon capture" actually means something, while at the same time not doing a damned thing to really make it work. Activists and others are right now championing ideas that were originally conceived by the very people and corporate entities that caused all this mess. Ideas we like because they allow us to keep living in the civilization we built and having all the cool tech and toys. If we just fix this problem or that issue, we can all keep our international air travel, our Amazon Prime, our nightly dining out, and even our air conditioning!

Awesome! Excellent work, everyone.

We are the ones let this cycle of idiocy define our future. And now it is really too late to do anything to change that. Sorry to be the one to shit in your soup, but that is just the way it is. The system in place is too big, and too entrenched, to stop now. There is not enough time remaining to really even give it a good try. Hell, I've only been writing this book for a few months, and things are coming apart so fast in the world that I keep having to go back and change things from "could happen soon" right to "so, this just happened." The snowball is rolling down the hill already, the feedback loops are looping, the tipping points are tipping, and the safe boundaries are in the rear-view mirror. We have corrupt leaders in place everywhere, parasitic sociopaths at the heads of our industries, and inept policy makers at every turn. They are not going anywhere anytime soon, certainly not soon enough, and those few who do are already slated to be replaced by more of the same. Every last one of them either oblivious or uncaring to the fact that the planet is going to be roasted and civilization is going to crumble along with it.

And so, here we are. There is all sorts of hopium out there, it's our newest drug of choice these days. From "carbon capture and storage" technology, which doesn't really work yet, to campaigns like "Net-Zero by 2050," as if we actually have that long. The truth is that hope is an illusion. Hope lets you continue going on with your day, doing your small part to work for the good, while those in power continue to do as they have always done, Business-As-Usual. That's all it takes to make people turn away from the looming problem, just a little hope that it will all get worked out in the end. And they carry that hope with them through the day, every day, seeing the problems but thinking "well, that sure is terrible, but we are all working to fix it!" And because of that, they won't see it coming. Not really. They won't be ready for the day when it all just falls apart, when the house of cards that is modern civilization topples flat before their eyes. As long as they have at least a shred of hope to hang onto, they will continue to try and fight to remain in a dying system, even if they die with it.

Don't be one of them. Don't embrace false hope. I am not saying you have to give in to despair and depression, far from it. Live your life. Prepare as you can and know what it really is that you are preparing for. Enjoy what we have right now, while we still have it. Don't waste what you have in the search after false hopes and corporate-sponsored dreams of a shining future. Live for now and get ready for the tough times ahead. Because there really is no going back to those old dreams. Time to make some new ones.

What comes after? Just what does the collapse of civilization look like? Well, that is a very long subject, with a great many potential answers. Possibly the subject for my next book. But one thing I know, its not going to be pretty. Knowing that, and embracing the truth of what is coming, is the very first step that must be taken in order to even think about effective preparation. Because we are not getting ready for some emergency that may happen, but for one that most certainly will. And it won't be like a tornado or a flood, or even like a war, after which there will be a return to something like normal again. There is no rebuilding after collapse. There is no insurance company to pay out for your ruined home, no construction company to hire for the rebuild, and no money with which to pay for it or a government to regulate the process for safety. There is you, hopefully alive, and those other people and things which have survived near you. Whatever is left is all there will be after that.

So, my advice? Get yourself and your family ready and stop wasting your time and effort in the search for a return to normal. It sucks, but it isn't coming back. Right now, it is as good as it will ever be.

I hope to see you out there, in the Wasteland.

ABOUT THE AUTHOR

"That's what I do. I drink, and I know things." – Tyrion Lannister

I am nobody special, not really. I am not a scientist or a researcher by education, not a policy maker of any sort, and I certainly hold very little power or influence in the world. I am only nominally in charge of the cats around here, and that is only by the consent of the governed. What I am, however, is a consumer of knowledge, usually just for the sake of it, because more than anything else in life I just want to understand things. Learning is what I value above all else, and for me there is no replacement for it. To strain the recently overused quote above close to its breaking point, it does indeed sum up my own character quite well, as does that of Tyrion himself. We share the same moral ambiguity, and I am the sort who is almost always overlooked and underestimated, which is quite by design. My politics are all over the place and tend to reside closest to pragmatism than to anything else. I am interested in facts, and I do not require myself to agree with them personally in order to recognize their validity.

For most of my life I have been something of an entrepreneur, but not of the conventional sort. I avoid things such as traditional employment like the plague. In fact, I've had covid more times than an actual job. If anything, I am more of a master of the side hustle than I am of any particular business pursuit, and that has taken me from the highest of tech to lowest of drudgery and back again many times. But such has given me all the time in the world to devote to activities of my own choice. In my recent years, much of my efforts have been spent learning about what is happening in the world and preparing myself to deal with the coming changes which I recognize as inevitable, a fact that has made me something of a buzzkill at parties.

Still, that should make me an immensely popular wasteland dweller after the collapse.